_____ 님께 드립니다.

해외도로공사 사업관리 1

공사관리업무
Managing Overseas Highway Project, Site Management and Control

초판 1쇄 발행 2025년 8월 20일

지은이 이관영
표지사진 신공항하이웨이주식회사
사진 서울고속도로주식회사
촬영 이대근
원고정리 박진하
펴낸이 장길수
펴낸곳 지식과감성#
출판등록 제2012-000081호

교정 정은솔
디자인 이현
편집 이현
검수 주경민
마케팅 김윤길

주소 서울시 금천구 벚꽃로298 대륭포스트타워6차 1212호
전화 070-4651-3730~4
팩스 070-4325-7006
이메일 ksbookup@naver.com
홈페이지 www.knsbookup.com

ISBN 979-11-392-2736-9(93530)
값 17,000원

- 이 책의 판권은 지은이와 지식과감성#에 있습니다.
- 이 책 내용의 전부 또는 일부를 재사용하려면 반드시 양측의 서면 동의를 받아야 합니다.
- 잘못된 책은 구입하신 곳에서 바꾸어 드립니다.

지식과감성#
홈페이지 바로가기

해외도로공사 사업관리 1
공사관리업무

Managing Overseas Highway Project,
Site Management and Control

목차

여는 글	6

1부 해외건설사업의 단계별 업무

1.1 정보 입수 및 입찰참가 결정	11
1.2 견적 및 입찰	14
1.3 계약	19
1.4 착공 준비	20
1.5 공사관리	23
1.6 준공	26

2부 Project Control Plan

2.1 Project Roles & Responsibilities	28
2.2 Local Engineer	32
2.3 Contract Administration	34
2.4 File System	41
2.5 Contract Registers	43
2.6 Contractual Correspondence	44
2.7 Contract Commencement	46
2.8 Issue of Drawings and Other Document	52
2.9 Subcontractors	56
2.10 Site Instructions	58
2.11 Daily Diaries and Report	59
2.12 Quality Control Field Inspection	62
2.13 Survey Checks	67
2.14 Site Meeting and Minutes	70
2.15 Monitoring of Progress	74
2.16 Progress Report	77

2.17 Financial Progress Monitoring	80
2.18 Measurement and Certification of Quantities	82
2.19 Interim Payment	94
2.20 Variations	96
2.21 Environmental Factors	102
2.22 Public and Relations; Health and Safety	105
2.23 Accident Report	107
2.24 Inspection Check Point	108
2.25 Submittal and Approvals	110

Appendix

A. Document Tracking Forms	122
B. Construction Supervision Forms	127
B.1 Correspondence/Communication Forms	128
B.2 Supervision Forms	135
C. Contract Administration Forms	158
D. Quality Control and Geotechnical Investigation Forms	184
E. Environmental Monitoring Forms	204

닫는 글	210

여는 글

건설 현장의 엔지니어는 공사 중 언제, 어디서, 누가, 무엇을, 어떻게, 왜 수행하는지를 확실히 파악해야 한다. 이는 도면과 기술시방서 기준에 따라 안전하게 시공하고 요구되는 품질을 확보할 수 있다. 계약서와 관련 법령은 이러한 책임과 의무를 엔지니어에게 분명히 요구하고 있으며 이것이 공사관리(Construction Management)와 사업관리(Project Management)의 핵심이다.

이 책은 국내 최초 민간 투자사업인 인천공항 고속도로 건설사업에서 사업 시행자로서 관계기관과 제정된 『사업관리 절차서(1996)』를 해외 현장에서 『Project Control Plan(2019)』으로 개정·적용한 공사 관리 사례를 정리한 것이다. 공사 중 업무 절차에 따라 어떻게 문서화하고 기록하며 운영하는지를 설명하였다.

해외현장은 사업관리업무를 '영어를 잘하느냐'가 업무 능력의 기준이 되곤 한다. 하지만 진정한 전문성은 해당 공사의 기술적 경험을 공유하고 협력하는 데 있다. 『Project Control Plan』은 이러한 소통을 위한 도구이자, 사업을 하나의 방향으로 통합하는 관리 수단이 될 수 있다.

제1부에서는 해외 건설사업의 정보분석부터 준공까지 프로젝트의 전체 흐름을 단계별로 정리하였다. 이는 사업 관리자들이 단계별 업무를 쉽게 이해하고 효과적으로 대응할 수 있다.

제2부는 『Project Control Plan』은 참여자의 역할, 책임, 주요 업무를 정리하고 실무에 활용할 수 있는 양식과 관리 문서를 함께 수록하였다. 이는 현장의 공사 관리 자료이며 나아가 준공 도서의 일부로도 활용될 수 있다.

앞으로 건설 사업관리가 표준화되어 Project Control System이 되면 엔지니어들이 PM의 주체가 되어 Project Manager로 활동할 수 있다. 이들은 발주기관으로부터 책임과 권한을 위임받아 기획부터 준공까지 전 과정을 관리하는 발주자의 파트너가 된다. 또한 공사 관리회사의 부분 별 엔지니어는 Construction Manager로서 발주자의 공정, 품질, 안전 등의 업무를 책임진다. 건설사업관리 업무가 체계적으로 되는 것이다.

여러분의 더 많은 자료가 표준화된다면 우리나라 건설 산업은 더 발전될 것입니다. 그동안 국내외 현장에서 함께 수고하신 모든 분께 깊은 감사의 마음을 드립니다.

2025년 8월

1부

해외건설사업의
단계별 업무

업무 흐름도

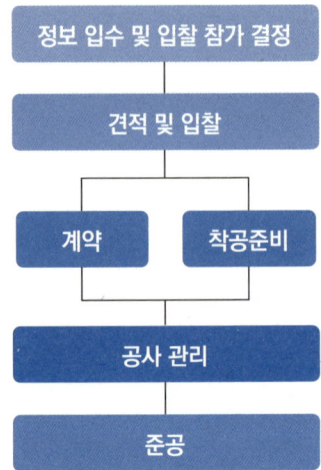

1.1 정보 입수 및 입찰참가 결정

발주 국가 관련 정보 입수

외국 업체 진출 여건
- 외국 업체의 영업활동에 관한 법규 및 제도
- Agent 관계법(대리인의 권한과 의무)
- 불법 중개인 금지규정 여부와 대응 방안
- Exclusive Agent 계약 주의, 중복된 대리인 계약 주의
- 유력 인사 및 적격 업체 파악
- 대리인 수수료율 및 지급방법, 계약 유효기간
- 출입국 관리제도
- VISA 발급 소요일수
- Work Permit 발급 절차 및 소요일수
- Regional Risk 파악(풍토, 정치, 경제, 외교 등)
- 경제개발계획-국토종합개발계획

금융제도와 세제
- 통화의 안정성
- 과실 송금의 규제
- 현지 금융제도와 금리 수준
- 각종 세제 및 특별 조항 발굴
  - 직접세: 법인세, 소득세, 사회보장세 등
  - 간접세: 부가가치세, 관세 등
- 외자 관계 법령 및 도입절차
- 현지 보험 제도

공사 계약 제도
- 입찰제도(공개경쟁, 지명경쟁)
- 지급 보증 제도
- 현지계약서 양식(국제표준계약서와 비교)
- 하도급 의무와 제한(현지 업체 의무 고용 방안)
- 선수금 및 기성 지급 조건
- 배상금 및 지체상금 등의 상한선 유무
- 중재제도 및 사법제도
- 하자보수 조건

현지 노동 조건
- 노동 관계 법규 및 근로자 입국 허가
- 노동조합 결성
- 근로 보험
- 임금 수준과 지급 방법
- 근로계약 조건(제3국인)
- 근로 시장의 현황
- 노동 생산성

기계 및 자재 사정
- 현지 조달 가능 품목과 가격
- 현지 기자재의 유통 구조
- 현지 기자재 공급의 안정성
- 운·수송 문제점
- 국산 및 외산 기자재 구입의 제약조건과 소요시간

현지 건설 시장의 동태
- 기 진출 외국 업체의 정보
- 현지 업체의 동태
- 기 진출 아국 업체의 정보
- 자국화 시책에 의한 외국 업체의 활동 여건 및 제한

사업별 및 발주처의 개발계획 입수

발주처의 개발 계획 및 공사 발주 정보
- 신규 개발계획
- 발주 계획(타당성 조사, 설계, 시공, 건설사업관리)
- 발주처의 계획 예산(단, 발주처 소문에 현혹되면 안 됨)
- 설계 용역회사 및 건설사업관리단 업체
- 발주 형태(공개경쟁, 지명경쟁, 단독 초청 등)
- 계약 형태(총액계약, 단가계약, 총액 및 단가 혼합계약 등)

입수된 공사 정보의 확인
- 신문 기사 및 관련 자료 추가 입수 검토
- 발주처와 용역업체를 통해 공사의 발주 진척 상황 파악
- 발주처 분위기와 담당자 파악
- 유력 인사 및 인적 네트워크 파악

외국 업체, 외국 정부, 국제기관의 동정
- 외국 업체 및 현지 업체의 수주 활동
- 자료 및 국제금융기관의 영향력
- 특정 국가에 대한 정치적 우대 경향
- 외국대사관의 수주 지원 동향
- 기 진출 업체의 영업 형태와 손익 현황

PQ 준비, 협력 및 지원 방안 모색

PQ 작성 및 제출
- 기 진출 분야에서의 전문성 및 특수 공법 소개
- 발주처의 자격 심사 기준 파악(설계 용역사 활용)
- 경쟁업체(PQ Brochure) 입수하여 전략 수립

공동 참여 모색 및 지원
- 아국 업체 간 공동 참여(JV, Consortium) 출혈경쟁 지양
- 대규모 공사 시 아국 업체 공동 참여로 과중한 Risk 경감
- 공동 참여 계약서, 공동 참여 PQ 준비
- 변호사 자문

국제적 협력 방안의 모색

경제 협력 관점에서의 차관 및 기술지원 동태 파악
- 국제 금융기관 및 선진국의 차관 공여, 기술지원 동태 파악
- 국제 금융기관에서 발간하는 정기 보고서 구독

국제적 협력 방안 모색
- 선진 업체와의 공동 참여 모색(JV, Consortium, 하도급)
- 현지 업체와 협력 방안 검토(합작법인 설립, 기술훈련 등)

합작 추진을 위한 사전 양해각서 작성
- 계약 당사자 확인
- 합작회사의 명의
- Sponsor, 합작회사의 대표 선임
- 각 당사자의 작업 구분
- 합작 추진 비용 및 손익 분담 방법
- 주요 인력
- 영업장소 및 주사무소
- Proposal 준비 및 Negotiation 시 발생 비용 분담
- 합작 사업 운용에 대한 초기 의도

합작 계약 체결 준비
- 계약 목적
- 사업 개시일
- 합작 사업 대표 및 Sponsor의 지정
- 합작회사 이사회 운영 방법
- 각 당사자의 책임과 의무
- 사업 추진을 위한 인력, 자재, 장비 등의 자비 부담에 관한 방법
- 자본 불입 및 경리 방법, 회계 연도 수립
- 손익 할당 및 분배
- 보험 관계
- 계약 종료, 계약서 수정 방법 및 절차
- 재무 기록 유지에 대한 책임
- 합작 계약 관련 합의 약정서 적용 범위
- 변제조건
- 등록 사항

하도급 계약 체결 준비
- 공사 금액 및 범위
- 책임과 의무 사항 정의
- 공사비 지급 방법
- 입찰 도서 및 현장 조건 숙지
- 소요 공기
- 공사 감독 및 시공자와 관계
- 배상 및 하자보수
- 안전 관리 및 보험 범위
- 적법성
- 공기지연
- 양도 및 파산
- 원청자의 자재공급

국제 변호사 및 현지 공인회계사 자문
- 각종 계약서
- 양해각서
- 합작법인 설립 추진
- 기타 향후 법률상 문제를 야기시킬 가능성이 있는 서류

참가의사 결정

공사 수행 능력 판단
- 동종 공사의 경험
- 현지 보유 인력의 전용 가능성
- 현지 및 인도 지역의 장비 전용 및 차용 가능성
- 기 수주하여 시공 중인 공사 금액
- 자금조달 방법
- 협력업체의 성실도
- 국내외 업체 간 경쟁 정도
- 예상 수익

1.2 견적 및 입찰

사전 조사와 자료 수집

입찰 도서의 구성
- 입찰초청서(Invitation for Bid)
- 입찰 유의 사항(Instruction to Bidders)
- 계약 일반조건(General Conditions of Contract)
- 계약 특수조건(Particular Conditions of Contract)
- 입찰서 양식(Form of Tender)
- 부록(Appendix)
- 발주처 공지 사항 통보(Clarification)
- 시방서(Specifications)
- 물량명세서(Bills of Quantities)
- 일위대가(Unit Price Analysis)
- 도면(Drawings)
- 발주처가 제공하는 공사 수행 시 필요 정보(Information & Data)

입찰 도서의 상세한 조사와 발주처의 유권해석
- 계약서 양식 및 조건 검토
- 계약 방식 및 조건
- 견적 조건
- 대안 설계의 유무
- 도면과 물량명세서를 상호 비교 후 누락 전 물량 조사
- 주요 지정 하청 공사의 공정계획
- 발주처 질문 사항에 대한 회신
- 적용되는 설계, 견적의 시방, 표준 등(예: AASHTO, BS, TEM 등)
- 기진출 업체 시공 단가
- 기타 정보 입수 단계에서 나열한 각 항목에 대한 재검토

현장 지형 조사
- 현장 및 인근의 기존 환경 도면
- 현장 구역과 경계선
- 가건물, 자재 야적장, 작업장
- 진입 방법과 도로, 울타리
- 사토 장소
- 부근 토지의 소유자
- 부지 권한 확인
- 용수원 확보
- 재 임차 현황
- 인근 가옥 주변 현황
- 지형의 고저
- 작업 구역과 통행금지 구역
- 매설 시설물, 가공선 등 유무

현장 및 인근의 지질조사
- 암반의 종류, 상태 및 지층 형성 조사 지점
- 지반의 경질 상태
- 기존 우물 및 지하수 위치와 깊이
- 토질 상태(연역 지반의 정도)
- 신축공사와 지반 작업의 윤곽
- 골재원의 위치와 공급 가능양

현장 및 인근 공공시설
- 급수 배수, 전력, 전화, 연료 소요량 및 용량, 현황과 조건
- 가설비물의 위치
- 확성기, 무선전화의 설치 유무
- 저장 방법, 용량 및 장소

현장 및 인근의 장비
- 장비 보유하고 있는 부품 재고 파악
- 수리 능력 및 부품 구입 가능성

입찰 참가 지명 접수
- 발주처로부터 입찰 초청장 접수, 공관에 보고 및 영사 확인
- 입찰 초청받은 타 업체 확인

입찰 도서 접수 및 현지 조사
- 발주처의 입찰 인수 및 현장에 대한 설명 및 질의
- 업체의 독자적 현지 조사

견적 전담 편성
- 공사 종류별 견적 전담 편성 및 자료 확인
- 공사 수행 시 임명된 PM을 견적에 참여

입찰 도서의 검토
- 입찰 도서 및 현장에 대한 발주처의 설명
- 현지 조사의 결과를 검토하여 견적 준비

현지 추가 조사
- 상기 검토 결과를 토대로 현지 재조사 실시 및 보완

질문 사항 제출과 회신 접수
- 검토 및 현지 추가 조사 결과로 질문 사항 작성 후 발주처 제출
- 발주처 회신 분석

입찰 도서 재검토
- 현지 재조사, 질문 회신, 추가 조건 등

상세설계 실시
- Turn-Key 공사의 경우 설계 실시

공사 물량의 산정
- 공사 설계서상의 물량에 의거하여 공사 종류 분류 (WBS)
- 시공상 필요한 물량 산정

시공 계획의 입안과 시공 방침 결정
- 시공의 기본 방침 결정
- 시공 방법
- 공정계획
- 가설계획
- 자재계획
- 장비계획
- 노무계획
- 외주계획
- 안전 위생 관리계획
- 공사 예산 작성
- 공사 관리 조직계획

일위대가표 작성
- 자재, 인력, 장비 등의 단가를 결정하고 이를 견적에 반영

입찰 도서의 검토
- 작업 단가의 내역을 작성
- 공사 종류, 단위 물량 당 자재, 인력, 장비비 산정
- 집계한 후 작업 단가 산출

가설 공사비 산출
- 직접 공사비와 같은 요령으로 산출

현장 경비 산출
- 현장 발생 총 경비를 계정과목별 집계

본사 지사 경비 및 이익 계상
- 현장에 부과되는 본사 지사 경비 및 적정 이익 계상

자금 이자 산출
- 공사 부담으로 해당되는 것 계상

견적 집계
- 직접 공사비, 가설 공사비, 현장 경비, 본지사 경비, 자금 이자 집계

종합 검토
- 견적 내용을 종합적으로 검토하고 필요시 보완 수정

입찰가격의 결정
- 단가계약 경우 간접비 배분 방법 결정
- 유사 시공 단가와 비교 검토
- 최종 제안 금액과 대안 검토 및 결정

해외 공사 견적의 특수 문제점과 대응 방안

계약조건 및 시방 등의 충실한 파악	CLAIM 대비	설계 용역사와 접촉하여 자료 입수 및 활용	현지 업체 및 근로자의 생산성 판단과 물가 임금 상승의 예측
• 입찰 도서에 기재된 모든 조건을 신중하게 분석, 검토	• 계약 후 조건 변경으로 인한 Claim 대상이 발생 시 즉시 제기 • 기간을 놓치면 능력 부족으로 취급 • 향후 Claim 제출을 위한 자료 작성에 사용 • 견적 자료와 견적 결과를 철저하게 정리하여 보관	• 타당성 조사 설계를 수행한 용역사의 자료 이용	• 현지 조사 및 견적 시에는 반드시 Escalation을 고려하여 결정

입찰 서류의 최종 검토와 주요 계약조건 재검토

지급 보증	착공지시서 및 현장 인수 조건	불가항력, 특별 위험, 제외된 위험, 공사 중지	지체 상금 (DELAY PENALTY)
• 보증의 유효기간 명시 • 무조건 지급보증(Unconditional Bond)의 지양 • 부분 준공 혹은 선수금 상환 시 보증 금액 감액 • 자재, 장비의 선급금을 지급 보증 대신 담보보증(Security) 대체	• 계약 체결 후 적기에 착공 지시나 현장 인수 • 전 현장 혹은 부분 현장 인수 인지 확인	• Regional Risk가 높은 국가인 경우에는 특히 국제 표준계약서의 본 조항과 비교 검토하도록 하되, 공사 중지의 서면 통지가 힘들 경우 시공사가 독자적으로 긴급행위를 취할 수 있도록 명시	• 발주국가 입찰 관계 법규에 지체상금에 관한 규정의 유무 확인 • 총 지체 상금액의 상한선을 계약서에 명시

하도급 의무와 특혜의 승계	공사 대금 지급	CLAIM	보험
• 지명 현지 업체의 조항의 경우에는 거부권을 확보, 특별계약 유도 • 원청자의 권리, 의무가 하도자에게도 승계되도록 계약서에 명시	• 적용 환율이 환차손을 받지 않도록 장치 • 현지화, 외화, 제3국화의 비율을 명시 • 현지금융 고려하면 공사 대금 등의 채권양도 금지 조항을 삭제	• 계약 변경을 유발하는 모든 사항에 관해 Claim을 제기할 권한과 충분한 행사 기간 및 적절한 Claim 보상 절차 명시	• 국제 정세 불안이 고조됨에 따라 Risk도 커지므로 보험 가입 • 계약조건상의 보험 조항 외 관련 현지 법령상의 규정도 비교검토 • 공사보험, 인명과 재산에 대한 보험, 제3자보험, 산재 보험 등

공기 연장	물가 연동(ESCALATION)	설계변경(VARIATION)	분쟁조정 및 해결
• 발주처 측의 귀책사유로 인한 경우 추가 비용 보상 청구권 확보 • 공기 연장과 관련되는 조항 점검 • 현장 인수, 도면 제공, 공사 중지, 추가 공사, 지체상금, 불가항력	• Negotiation조건 협의, 시기, 기준 물가지수 등 • 본 조항 발효 시점에 증빙자료 제출을 통해 Escalation 확보	• 일정 %를 초과하거나 미달하는 경우 • 정식 서면에 의한 변경 지시가 아닌 경우 -해석 상의 변경인 경우 • 시공회사 측이 보상 청구권을 확보하도록 명시 • Cost & Profit 모두 반영되도록 추진	• 보통 발주국가 중재법과 국제상사조정(ICC) 등의 중재 고려 • 하도급 계약 체결 시 중재 판결은 구속한다는 것을 명시

입찰 서류 준비 및 작성

확인 사항
- 발주국의 입찰 관계 법규를 숙지해 입찰서 작성 및 입찰방법 검토
- 입찰 유의서 의거 발주처가 요구사항에 부합되도록 입찰서 작성
- 발급된 Clarification 또는 Addendum을 체계적으로 정리 반영
- 입찰 서류 오자 수정 방법을 충실히 이행
- 입찰 서류에 서명, 날인 방법과 서명자의 합법적 지정
- 입찰 서류 제출 부수 확인
- 입찰 서류 포장, 제출 방법 및 제출 장소 확인
- 입찰 서류 제출 마감 시간 준수
- Bid Security 발급 소요 시간을 고려하여 시간적 여유 확보 필요
- 담당자 발주처 직접 방문, 우편, 전자우편 등

보안 유지
- 지사 또는 현지 출장 인력과 교신 시, 입찰 관련 정보 보안 유지
- 입찰서 항공 운반 시, 수화물 처리 지양

BID SECURITY 제출

입찰 보증서 문안 검토 → **보증기간 만료일 전 발주처가 SECURITY 유효기간 연장을 요청 조치** → **계약 이행 보증 (PERFORMANCE SECURITY)을 제출 엄수** → **수의계약의 경우에는 입찰보증 제출하지 않음**

- 입찰 유효기간과 1회 Security 유효기간 대조 확인
- 발주처 몰수 요건 확인
- 입찰 서류에서 요구하는 보증 금액 확인
- 필요시 복보증 제출
- 입찰 서류에서 명시한 보증 상대처의 확인

계약 협상팀 편성

NEGOTIATION TEAM 편성
- 견적 담당자 및 PM
- 수주 담당자
- 계약 담당자
- 외국어에 능통하고 발주국에 대한 이해가 있는 자

계약 협상을 위한 질의·토의 실시
- 견적 시 이용한 자료 및 견적 과정을 정리한 자료 준비
- 입찰 후 입수된 타 경쟁업체의 입찰 금액 동태 파악
- 모의 협상 실시

발주처 및 경쟁업체 동태 파악
- 모의 협상 시, 발주처 특성 파악
- 입찰 서류 개봉 후 발주처 동태 수시 점검
- 입찰 후 입수된 타 경쟁업체의 입찰 서류 및 동태 파악

낙찰의향서 접수

- 낙찰의향서(Letter of Intent)는 계약 체결을 위하여 발급하는 낙찰통지서(Letter of Acceptance)와 달리 계약 협상 초청임을 유의
- 낙찰의향서를 접수한 후 공사 착수 준비를 하더라도 추후에 낙찰통지서를 받지 못할 경우 그 비용은 보상받지 못함에 유의

계약 협상 시 유의 사항

계약 협상 상대방의 동기 파악
- 평상시 발주처를 자주 접촉하여 주요 관심사를 파악하여 활용
- 발주처는 주로 가격 삭감에, 입찰자는 계약조건의 변경에 주력할 것이므로 이 사항들에 대한 충분하고 설득력 있는 자료 준비 계약 협상 회의 참석 예정자의 인적 사항 파악

계약 협상 기법 발휘
- 개인적인 자질 위에 몇 가지 원칙을 응용하여 회의 대처 능력 발휘
- 지나친 전문용어 남발 금지
- "Give & Take" 원칙을 명시하여 상대방의 입장을 고려
- 발주처의 이익을 충분히 이해시킴
- 협상을 위한 발언에 초점 설정
- 주도권을 잃지 않도록 요구사항 별 확답을 얻도록 노력
- 거부 의사를 표시하더라도 설득력 있는 자료 통해 타당성을 규명
- 회의를 통해 협상을 마무리하기 보다는 추가 협상 가능성이 좋음

1.3 계약

낙찰통지서 접수

- 낙찰통지서 발급 발주처 측과 협의
- 계약 체결 시 제출할 계약이행보증 등 고려 체결일 협의

계약이행보증 및 선수금 환급보증 제출

계약이행보증 및 선수금 환급 보증의 문안 검토	계약 서명을 위한 위임장 발급	계약서(CONTRACT AGREEMENT) 구성	계약서 작성 및 서명
• 공기, 하자보수 기간과 1회 Security 유효기간의 대조 확인 • 무조건 지급보증 (Unconditional Bond) 지양 • 부분 준공 혹은 선수금 상환 시, 보증 금액 감액 • 양도금지 명기 • 발주처 측 요구사항, 위험 부담을 보험으로 분산 조치	• 적법한 위임자 지정 • 적법한 절차에 따라 공증 및 필요한 인증 후, 제출	• 계약일반조건(General Conditions of Contract) • 계약특수조건(Particular Conditions of Contract) • 시방서(Specifications) • 도면(Drawings) • 부속서류 및 계약(Related Documents/Agreement) • 입찰서(Bid Proposal) • 공사이행보증 (Performance Security) • 부록, 조건 변경 등 • 통지된 사항 Clarification, Addendum 확인	• 입찰서에 포함된 표준 계약 견본을 참고하여 협상 합의 사항 반영 • 발주처와 주고받은 공문을 부속서류에 포함 • 정식계약과 위배되는 공문 이나 문서는 무효임을 명기 하거나 배제

계약 체결 후, 조치 사항

ENTITY 등록	기타 등록 및 가입 업무	현지 법률 자문 및 세무/회계 협력업체 조사
• 현지 외국 회사법에 따라 Project Office, 지사 또는 법인을 설립 • Tax Identification Number 발급	• 상공회의소(Chamber of Commerce Industry) 가입 • 기술자V등록(VISA 및 Work Permit 발급)	

1.4 착공 준비

현장 조직의 편성

현장대리인(PROJECT MANAGER) 임명
- 당해 공사에 적합한 조직, 지휘, 인사, 장비, 자재, 공법, 공정, 품질, 안전, 자금, 행정 등에 관한 능력과 책임감이 있는 자
- 동종 공사 기술 지식과 시공 경험자
- 외국어 능통한 자
- 노사협조체를 통솔할 수 있고 직원 복지에 관심이 있는 자
- 가능한 해당 공사의 견적에 참여한 자로서 준공까지 가능한 자
- 발주처의 요구조건에 부합하는 자

현장 담당직원의 자질
- 현장 조직 편성과 소요 인원 산정
- 해당 분야에서 전문 지식, 기술 관리 능력, 현장 경험을 가진 자
- 공사 수행에 요구되는 외국어 능통자
- 제3국인을 통솔할 수 있으며 신뢰를 받을 수 있는 자
- 업무에 충실하고 책임감이 강한 자

현장 인수 및 측량

현장 인수
- 선발대 현장 도착
- Notice to Proceed(NTP/착수지시서) 접수
- 건설사업관리단 현장 도착 확인 및 착공 준비 협의
- 현장 합동 조사 실시 후 서면으로 현장 인수(현장 인수 불가 시, 그 이유를 서면으로 발주처 제출)
- 현장 정밀 조사 실시
  - 지형
  - 지질, 지하수질
  - 기상
  - 동력, 용수, 통신
  - 수송망
  - 철거 대상물

시공 측량
- 측량성과 자료, Temporary Bench Mark(TBM/임시 수준점) 인수
- 실측 결과를 인수한 성과자료와 비교 검토
- 시공 측점 표지 설치

공사 계획과 실행예산의 편성 1

공사 계획과 공사 부수 계획 수립	가설 계획	인력 투입 계획	자재 투입 계획
• 시공 방법은 여러 대안을 가지고 충분히 검토하여 공사 계획 작성 • 시공 순서를 적정하게 배치한 Critical Path Method 작성 • 각 작업별 소정 공사 기간을 적정하게 배정 • 인력, 자재, 장비, 자금의 수급 가능한 공사 부수 계획 작성 • 전체 및 공구별 공기와 하도급 공기가 계약 공기 대비 적정 작성 • 자금 수급 계획, 기성 신청 계획 및 실행예산 집행 가능 계획 • 발주처 요구조건을 감안하여 공사 계획 작성	• 가설물의 평면 배치도 고려 사항 - 사무실, 숙식시설, 자재저장 시설, 장비 주차 등 제시설 관리 - 인력, 장비의 왕래에 지장이 없도록 고려 - 여러 대안 작성 후, 최종안 결정 - 변경에 대비한 융통성 부여 - Camp 설치 계획서 작성 - 가설 계획이 결정되면 설계도와 내역서를 작성하고 공사비 판단 • 제반시설의 설치 계획 시 관리 사항 - 제반 가설물의 사용 재료, 규격, 구조 소요량과 비용 산정 - 현장 사무소의 수용 인원, 사용 기간, 설비의 산정 - 창고, 재료 저장의 저장물의 종류, 저장량, 저장기간 산정 - 동력, 용수, 통신 등의 소요량 산정 • 장비 기구 설치계획 시 고려 사항 - 공사의 경제성과 공정에 맞는 종류 선정 - 성능, 조립, 설치 장소, 비용, 설치 방법의 연구와 설치 소요일수 - 필요한 부속 및 소모품 구입 가능성과 가격 - 검사 등의 후속 조치 • 발판비계의 설치 계획 시 고려 사항 - 사용 재료, 규격, 구조, 예산 관계 검토 • 가설자재, 기계기구의 운송계획 시 고려 사항 - 공정에 따른 반입계획 점검 - 운반조건 재확인(수량, 중량, 용적, 경비, 소요일) - 일시 적재나 반복 교통수단의 번거로움을 피함 - 장해물 운송의 인가 수속 조치 (관계부처 협의)	• 공사계획에 의거한 작업 단계별 노무 투입 계획 수립 • 적정 작업 배치 고려 • 아국, 제3국, 현지 기능공의 작업 능력과 생산성 파악 • 기능공 작업 지도 및 훈련 계획 • 시공 지도 요원의 경험을 고려, 사전 교육은 충분한지 점검	• 공사 계획에 의거한 자재 투입 계획 수립 • 발주기관 견본 승인 일정 계획의 작성 • 자재 명세서(Bills of Quantities)의 작성 - 구매 자재, 지정 자재, 전용 자재, 발주처 • 공급 자재 • 자재 승인 후 현장으로 반입 소요 기간을 고려한 계획 • 수입 자재원과 물가를 사전 조사하여 자재원을 선정 • 현지조달 가능 자재에 대한 사전 시장조사 • 전용 가능한 자재는 균일한 품질을 유지할 수 있는 제품 선정 • 주요 자재의 가격변동, 수급 동향 수시 파악 • 구입량, 구입 시기 결정에서 다음 사항 점검 - 일시 또는 분할 구매가 소요되는 자재 - 창고, 야적장의 준비 상태 - 생산지 및 운반 경로 - 특별 제작, 특수 자재인 경우 제작 회사 • 및 제작 기간 - 시작, 시험을 요하는 품목 확인 - 발주처 승인에 소요되는 시간 및 검수자와 감독관의 협조 • 규격의 정확성 검사

공사 계획과 실행예산의 편성 2

장비 투입 계획	외주 시행 계획	자금 수급 계획	실행예산 편성 시 고려 사항
• 공사 계획에 의거한 작업 단계별 장비 투입 계획 수립 • 보유 장비를 최대한 전용 또는 활용 • 장비에 대한 신규 구매 투자 억제 • 소요 장비의 가동률을 최대한으로 높이도록 장비 할당 계획 수립 • 임차 장비와 구매 장비의 경제성, 활용성 비교 검토 • 작업 범위를 고려한 장비 선정 • 필요한 부속 및 소모품 구입 범위와 그 가격 • 현장 및 인근 수리시설 확인 • 장비의 주기적 정비 계획 수립	• 현지 시공 능력 부족 또는 동원 소요 비용 과다 및 공사 기간 고려 • 하도급자 선정은 다음 사항을 고려하여 결정 -법인의 영업허가 소지 여부 -재무 건전 여부 -자금 가용 능력 -신용도 및 평판 -기술 수준과 공사 실적 -관련 공사 협조 • 명확한 협의를 통해 하도급 계약 체결 • 하도급 공사 계획은 본 공사 계획과 일치	• 자금조달 및 운용계획은 공사 계획의 공정(율)률에 따라 계획 • 특히, 인력, 자재, 장비 투입에 합치된 투입 계획 • 자금조달과 운용은 시공자의 자금 수급 능력에 근거	• 실행예산은 견적 참여자, 현장 담당자, 예산 담당자 협의 편성 • Negotiation 시 변경된 내용은 실행예산에 반영 • 공사 종류별 원가 항목 구분 • 물가 연동(Escalation) 조항 유무에 따라 검토 • 잡부, 운반, 뒷손질 등의 비용 개별 산출하여 필요 경비 명시 • M2 단가나 M3 단가가 아닌 각 집입 원가를 파악 • 구입량, 단가만이 아닌 각 반입 시기 방법 검토 • 기능공은 언제, 어디서, 몇 명이 필요한지 세부 산정 • 장비 투입 계획 구체적 작성 • 감독 지시 사항에 순응하고 약속은 성실하게 지켜 신뢰 구축 • 공사감독자가 합리적이지 않거나 횡포를 부릴 경우, 그때마다 입증 자료를 수집하여 차상급 감독 측에 제기하여 개선 시도(개선이 안 될 경우 발주처에 통보하여 공사 방해 요인 제거) • 공사에서 이익을 낼 수 있는 요인과 난점을 정확하게 파악하고 그 대처 방안 다각 검토 • 세부 내역 정리 및 유지 • 타 동종 공사의 실행예산과 비교하여 실편성 검토 • 가설공사(현지 사무소, 감독관 사무소, 하도급자 사무소, 숙소, 식당, 복지시설, 진료소, 장비 정비고, 창고 등) • 동원 공사 계획, 설계도면, 내역서, 인력, 장비, 자재 등 확인 • 일일 세부 작업 계획 작성 실시 • 작업 조 편성(시공 지도 기사와 기능 인력) • 일일 작업 목표량 지시, 장비 투입 지시, 자재 출고 지시 • 일일 작업성과 결산 • 모든 계약 이행 사항은 현장에서 착공부터 준공까지 서면 관리 • 감독 측과 회의 사항은 회의록 작성 유지 • 각종 장애물 이전 및 각종 표식판 부착 • 기공식 준비 • 발주처 및 관계 기관에 착공 보고 • 관계 기관의 공사 허가 획득 • 인근 주민 또는 민간단체에 대한 이해 협조 요청 • 토지, 건물 임차에 관한 계약 체결 • 보험 가입 • 현장 공사 행정 착수, 문서 수신 발신 기록 유지

1.5 공사관리

공사관리 1

현장 조직 관리	공정 관리	작업 관리	인력 관리
• 현장 조직에 따라 직원 확보 • 공사 수행상 현장 조직의 참모 편성, 지원 부서, 공구 편성 점검 • 현장 업무는 부서 편제에 적정하게 배치되었는지 점검 • 업무 지시는 적시에 전달되어 원만히 수행되는지 점검 • 부서 간 업무 협조는 원활한지 점검 • 구성원의 업무상 사기 격려	• 공사 계획표와 CPM 공정표는 PM의 주도하에 가능한 계획으로 수립 • CPM에 의한 세부 공사 계획 작성 • 설계도, 시공도면을 검토하여 실행 대비 계획 검토 • 세부 공사 계획에 따른 인력, 자재, 장비 계획이 가능하도록 점검 • 공사 종류별 공사 순서와 상호 간섭 유무 점검 • 공사 계획 대비 실적을 검토하여 공정 수정 및 만회 계획 대책 강구 • CPM 및 관리 업무 전산화 • 계약 특수 공법에 대한 기술 숙달로 공기 단축 차질 예방 • 기술팀 지원으로 조속한 시공 설계도면 승인 획득 • 건설사업관리단의 횡포에 대처하고 Claim 즉시 대처하여 불이익 예방 • 현장 기사 및 관리층, 건설사업관리단 사이의 정확한 의사소통 • 현장 관리 요원 및 기능공의 사기 높이고 이직 억제 • 공기 연장 관련 사항에 대한 Claim 제기 및 자료 유지 • 외주 관리 시, 단위 공사 종류 연관성이 저해되지 않게 관리 철저 • 합작 계약 시, 실공사 협조 관련 책임과 의무를 상세히 규정 • 공정에 따른 인력, 자재, 장비의 수급 점검	• 세부 공사 계획에 의거한 일일 작업 계획 작성 • 일일 작업 목표량을 제시 • 작업 조 편성과 장비, 자재, 용수, 연료, 부수자재 등 확인 • 작업 개시 전, 작업 내용 및 수행 방법 등에 관해 Orientation 실시 • 시공 지도 기사의 적절한 배치와 작업 감독 점검 • 일일 투입 원가와 시공 생산성을 비교 손실 대비책 강구 • 작업 중 규칙, 주의, 각종 표지판, 안전 장구, 안전 설비 점검 • 공구와 작업장 설비의 완비 후, 작업 실시 • 설계도와 시방 기준에 의한 작업 기능도와 품질 점검 • 작업 시간의 엄수 • 일일 작업 완료 후, 시공 생산고의 검측 확인 • 작업 방법의 개선 노력 • 작업 인력의 시공 생산고에 따른 성과급 지급 적정화 • 감독 측 작업 지시에 대한 이행 상태와 문제 발생 시 대책 강구 여부 • 하도급 시공 시, 작업 감독 계획 수립 및 감독 실시 • 공사 일지 기록 유지	• 공사 계획에 따른 인력 투입 계획의 공정별, 직종별 인원 수 조치 • 현장에서 인력 소요 계획은 향후 6개월간 매월 소요를 매월 말 제출 • 기능도 측정에 의한 선발로 생산성 제고 • 기능공의 직종별 작업에 적합하도록 작업 배치 • 노사 협의회를 설치하고 노사관계의 개선 • 노무 담당자와 공부 담당자 간의 업무 협의 원활화 • 현장 인력 투입에 따른 사전 문제점 검토 및 대책 수립 • 인력의 타 현장 전용 계획 수립 • 현지 취업 VISA 발급 절차를 사전 점검하여 인력 송출 대처 • 제3국 기능공 활용으로 원가 절감 • 취업자 등록 작성 및 출역 점검 체제 수립

공사관리 2

자재 관리	장비 관리	품질 관리	안전 관리
• 자재 명세서 작성 • 자재 견본에 대한 감독 측의 서면 승인 여부 확인 • 과오 주문 및 이중 주문 방지 • 자재 조달은 구매 계획에 의거하여 적기에 조치되는지 점검 • 자재원과 물가 조사 철저 • 자재 승인 전, 자재 가격 결정 • 자재 품질 및 규격이 시방 기준과 일치하는지 점검 • 자재 발주 시 공급자 선정은 생산능력, 신용도를 검토 후, 결정 • 자재 발주 계약은 조건을 명기하여 문제나 분쟁 방지 • 자재 발주 후, 생산량과 품질 관리 점검 • 자재 공급자는 승인 받을 때까지 운송, 취급, 파손에 대한 책임 명시 • 기자재 조작에 관한 설명서 및 보험보증서를 받고 보존 유지 • 자재의 재고 관리 체제 확인 • 자재 전용, 폐자재 활용 방안 수립 • 주요 자재의 가격 및 수급 동향을 수시 파악 • 사내 구매 절차는 원활하게 운영되는지 점검 • 부적합한 자재에 대한 처리 방안 수립 • 자재 정보의 수시 점검으로 구매처 변경에 의한 원가 절감 유도 • 공정에 따른 자재 적기 투입	• 장비 투입 계획대로 장비 투입과 배치가 조치되는지 점검 • 장비의 기종, 규격, 성능은 작업에 적합한지 구매 전 검토 • 장비 부서와 공무 부서 간 업무 협의가 원활하도록 조정 • 예방정비(Preventive Maintenance)는 계획을 세워 정기적 시행 • 회사별로 현지에서 정비 관리하여 현장은 임차 형식으로 장비 활용 • 현장에서 장비 직영 시, 장비 관리 부서의 운영은 독립채산제로 운영 • 장비 운행 일지와 정비 일지 기록 유지 • 장비 부품 조달원 확보	• 품질 관리는 설계도와 시방서에 기준 관리 철저 • 시방 기준에 의한 재질의 품질 보장 • 각종 품질 시험을 위한 설비(장비) 자료 구비 • 시험 요원의 인원수와 자질의 적정화 • 현장 관리 시험은 시방에 따라 실시 • 완성된 공사의 품질 보전을 위한 대책 • 품질 업무와 정보 교환으로 품질 향상과 재시공 및 하자 방지 • 자재 견본 비치실을 유지하여 투입 자재 대조 확인	• 안전 규칙, 안전 장구, 안전 설비, 안전 복장 완비 • 안전 관리 계획을 작성하여 정기적인 안전 행사 실시 • 안전사고 사례집을 송부하여 안전 교육 실시 • 일일 작업 착수 전, 안전 교육과 안전 점검 • 현장 내, 안전 설비의 완벽화를 위한 수시 및 정기 점검 • 기능공의 복장은 기직종 작업에 적절하도록 조치 • 자동차의 현장 내 속도 제한, 적재량 준수, 낙하 방지 조치 • 현장 내 구급, 의무시설 완비 • 사고 처리 요령 수립 • 안전 일지 기록 유지

공사관리 3

원가 관리	기성 관리	자금 관리	건설사업관리단 협조 체제
• 실행예산 대비 투입 원가 비교 분석 정기 실시 • 일일 작업 투입 원가의 적정성을 검토하여 원가 관리 개선 • 실행예산서가 실제 집행 대비 변경 승인	• Cost Engineer를 양성하여 계약 관리, 기성 관리 전담 • 현장소장 및 현장 직원들은 계약 내용을 숙지 • 기성 신청 계획이 공사 계획 및 공정에 맞춰 실행 가능한지 점검 • 기성 신청 후, 계약조건대로 기성 증명이 적시 발급되는지 확인 • 발주처 회계 부서 계통으로 기성 지불이 원활한지 확인 • Claim 사유 발생 시마다 계약조건에 명시된 기한 내, 입증 자료 • 제기된 Claim은 매회 기성에 일괄 청구 조치	• 공사 자금 수급 계획이 자금 조달과 운용에 차질 없이 적정한지 검토 • 기성 지불이 원활하여 자금 수급 계획의 집행과 합치하는지 점검 • 경비 지출이 적정한 통제가 되고 있는지 점검 • 자금조달과 운용 계획의 적정성과 절차, 사후 처리 사항 점검 • 자금조달은 자사 내에서 가용하도록 노력하며 외부 차입 억제 • 공사 추진에 지장이 있을 정도의 해당 현장 자금의 지출 억제 • 현지 세무 처리 관계 조사 및 대책 수립 • 현지 관계 법규 숙지 • 현지 회계사/회계법인 활용 • 절세 방안 강구 • 공동 참여(JV 또는 Consortium) 시, 현지 세무신고 방안 공동 모색 • 본지사의 관리하에 재무 분석 정기 실시	• 공사 계획과 실시에 관한 협조와 공사 계획표 제출과 공정 보고 • 승인도면과 시방서 인수 • 자재 견본 승인 계획 협의 • 시공 검사와 감독 요령에 대한 협의와 서면 승인 또는 합격 협의

1.6 준공

예비준공검사 및 정산
- 시공사는 준공 1개월 전, 준공 검사원 건설사업관리단 제출
- 감리단의 예비 준공검사 실시 결과 지적 사항 시정
- 예비 준공증명서(Taking-Over Certificate) 발급 접수
- 예비 준공증명서 발급일로부터 하자보수 기간 산정
- 대 본사 지사 준공 보고
- 관련 기관 준공 보고, 현지 유관 기관 준공 보고
- 설계 변경의 최종 정산 및 잔액 수령
- 유보 보증금 50% 환수 또는 Performance Security 기간 연장
- 현지 관청에 공사 완료 보고 및 인허가 행정 마무리
- 검측 항목에 대한 물량 정산은 증감 모두 반영
- 최종 정산 과정에서 누락된 청수 사항 재검토
- 발주처에 제출할 최종 정산서 검토
- 사내 정산은 실행예산과 대비하여 검토하여 실행 결과 보고
- 준공증명서 발급(Performance Certificate)

준공 도서 및 사용 기자재 견본 확보
- As-Built Drawing의 제 규격, 수치 등은 사전에 충분한 기간 여유를 두고 발주처 및 현지 관계 부처와 협의하여 추진
- 계약서류에 명시된 제출 준공 도서의 규격, 종류 및 부수 확인
- Maintenance Manual을 위한 자재 열람표는 자재 구입 시 점검
- 공사에서 기사용된 기자재의 견본 확보

2부

Project
Control Plan

2.1 Project Roles & Responsibilities

International Team Leader (TL)

The TL manages the overall project administration, ensuring the timely and quality execution of the construction work. He liaises directly with the Employer to uphold contractual agreements and meets the project's objectives.

Key Responsibilities
- Oversee the entire project and administer the construction contract.
- Lead and manage the Consultant staff.
- Implement a computer-aided contract management system for monitoring.
- Assist with resolving contractor claims and review payment certificates.
- Conduct regular reviews of the project's schedule and financial status.
- Promote teamwork and collaboration between the Contractor and Consultant teams.
- Liaise with the Employer on the project's progress and compliance.

Deputy Team Leader (DTL)

The DTL supports the TL, stepping in for the TL when needed, and manages the daily operations to ensure compliance with project timelines and quality standards.

Key Responsibilities:
- Act as the TL in their absence.
- Monitor progress and ensure adherence to contractual requirements.
- Attend construction progress meetings and manage technical approvals.
- Oversee preparation of regular reports and certifications.
- Ensure compliance with environmental and safety policies

International Contracts Specialist

Responsible for managing the project's contractual aspects, ensuring compliance, and addressing claims and variations.

Key Responsibilities:
- Maintain and update claims and variations registers.
- Assess and advise on the validity of contractor claims.
- Supervise the local Contract Specialist and support Consultant staff in contractual matters.
- Draft and manage correspondence on contractual issues.

International Highway/Pavement Engineer

Oversees the design and construction of highways and pavements, ensuring the contractor's compliance with technical and quality standards.

Key Responsibilities:
- Review pavement designs and recommend improvements.
- Supervise local engineers and surveyors during construction.
- Coordinate materials testing with the Quality Control team.
- Assist in monthly reporting.

International Quality Control Specialist

Ensures the quality and standards of materials and construction work throughout the project.

Key Responsibilities:
- Inspect materials and testing facilities.
- Monitor compliance with material sourcing and quality assurance processes.
- Provide guidance on testing and material approvals.
- Conduct audits of testing procedures and reporting.

International Bridge Engineer

Specializes in the design, review, and construction of bridge structures, ensuring compliance with engineering standards and contract specifications.

Key Responsibilities:
- Supervise bridge design and construction processes.
- Coordinate with the Quality Control Engineer to ensure material quality.
- Assist in monthly reporting on bridge construction progress.

International Geotechnical Engineer

Provides expertise in geotechnical matters, ensuring that construction methods and designs are suitable for the geological conditions.

Key Responsibilities:
- Evaluate contractor-proposed construction methods.
- Supervise geotechnical investigations and testing.
- Advise on compliance with geotechnical design and construction methods.

International Tunnel Engineer

Leads the design and construction of tunnel structures, ensuring compliance with safety, environmental, and technical requirements.
Key Responsibilities:
- Review tunnel design and excavation methods.
- Ensure compliance with environmental and safety standards.
- Supervise local engineers and provide tunnel-specific training.

International Environmental Specialist

Oversees environmental compliance, ensuring that the project adheres to environmental laws and best practices.

Key Responsibilities:
- Supervise the local Environmental Specialist.
- Prepare environmental management plans and conduct site audits.
- Liaise with the Employer and assist in environmental reporting.

2.2 Local Engineer

Highway, Bridge, Tunnel Engineers

Support their respective international specialists in daily technical supervision, ensuring the construction adheres to designs, safety standards, and quality controls.

Key Responsibilities:
- Assist in checking the setting out of works and verifying compliance with designs.
- Manage the measurement of work and inspect quality.
- Ensure the Contractor complies with safety and environmental policies.

Surveyors

Responsible for ensuring that survey control points are established and maintained throughout the project.

Key Responsibilities:
- Assist in setting out permanent works and joint surveys with the Contractor.
- Maintain survey records and control equipment.
- Provide technical survey support to senior engineers.

Quality Assurance Engineer

Supports the Quality Control Specialist in overseeing the quality of materials and workmanship.

Key Responsibilities:
- Conduct detailed inspections and material testing.
- Ensure all testing is documented and meets contract specifications.

Road Safety Engineer (Intermittent)

Focuses on road safety audits and inspections, recommending measures to improve safety.

Key Responsibilities:
- Conduct road safety audits at key project stages.
- Inspect completed works to ensure safety before public use.

Office Manager

Manages administrative tasks, including staff, documentation, and logistics, to support the smooth operation of the project office.

Key Responsibilities:
- Oversee local administrative staff and office operations.
- Manage travel, visa, and permit arrangements for international staff.
- Maintain financial records, correspondence, and filing systems.
- Procure and manage office supplies.

2.3 Contract Administration

General

The three principal entities in a construction contract are:
- The Client, referred to in the General Conditions of Contract as the Employer, who initiates, pays for and is the ultimate owner of the project;
- The Contractor, who carries out the actual construction;
- The Consultant, who administers and supervises the contract.
- The Consultant has two main goals with respect to construction contracts, both equally important, and one not to be achieved at the expense of the other:
- Progress
- Quality

Basic Concepts

The Engineer has to:
- Manage the Consultant's team;
- Make technical decisions as and when required;
- Certify payments to the Contractor
- Manage the contractual relationship with the Contractor;
- Assist the Client in procurement and managing the project, and act as his agent;
- Act independently of both Contractor and Client as valuer, certified and adjudicator.

The Engineer may delegate parts of his authority and duties

under the construction contract, if approval to do so is given by the Employer. A copy of the delegated powers is to be provided to the Employer and the Contractor.

Conditions of Contract

The Conditions of Contract set out the rights and obligations of the contracting parties – the Employer and the Contractor. These may be varied during the currency of the contract providing there is agreement by both parties. Conditions of Contract cannot be varied by the Engineer or the Consultant.

Specifications

The Technical Specifications are generally performance specifications in that they specify the end result of the work and make the contractor fully responsible for the correction of defective work without additional payment.

Technical Specifications cannot be varied by the Engineer or the Consultant. The Engineer sometimes has the obligation to approve the Contractor's methodology but the ultimate choice of method remains with the Contractor. The Engineer may reject all work which fails to meet the standards specified, regardless of any approval of the Contractor's methods and procedures.

Operational Matters Survey and Setting Out

The Employer is responsible for providing the Contractor with initial survey reference points. The Contractor is then responsible for establishing survey control for his own use and for the day-to-day setting out. Suitable site procedures must be established for checking the Contractor's setting out in key areas to ensure that the work meets the lines, levels, dimensions and tolerances required. (Refer to Procedure CA13 - Survey Checks).

A joint survey must be undertaken by surveyors from the Consultant and the Contractor before work commences to ensure that there is complete agreement on base data, and therefore earthwork quantities can be agreed without dispute during the currency of the contract.

Approval of Materials

The Engineer has the right to request test certificates, manufacturer's warranties and other evidence that the materials supplied comply with the Technical Specifications. Substitute materials are permitted where the contract so allows, subject to the Engineer's approval.

Acceptance Testing

Where the Specifications require materials to be supplied, or work to be performed, to specified standards or codes, acceptance must be based on tests prescribed in the codes and on the acceptance criteria stated therein.

Manufactured components or materials supplied by the contractor to be installed in the permanent work, whether inspected off-site or not, are subject to acceptance at the site.

Quality control methods are set out in Procedure CA12 - Quality Control - Field Inspection. Quality control checklists and an Audit Schedule have been prepared and are included in the Construction Quality Control Manual.

Engineer's Approvals and Directions

All work under the contract is required to be performed in accordance with the Specifications to the satisfaction of the Engineer. The contract also sets out those matters which require the Engineer's approval and those which require directions from the Engineer.

The Engineer is required to give consent or approval to various submissions made by the Contractor in accordance with the contract. Verbal consent or approval must be confirmed in writing and written consent or approvals must state:

"Consent or approval does not relieve the Contractor of responsibility for correctness of detail nor does it waive any of the Contractor's obligations under the Contract unless specifically stated in writing by the Engineer or the Employer."

Where work is subject to directions from the Engineer, responsibility must be accepted for the outcome of those directions and the contract could allow additional payment. All directions from the Engineer must state what work is to be performed and the method of payment, e.g. without additional cost or at a price to be submitted by the Contractor and agreed by the Engineer. (Refer to Procedure CA20 - Variations).

A direction from the Engineer may be issued:

- By letter to:
  - give notice to commence
  - delay the work
  - approve drawings
  - direct the Contractor to comply with the requirements of the contract
  - amend or clarify a detail on the Drawings or Specifications
  - vary the scope of work, etc.
- By Site Instructions (see Procedure CA10 - Site Instructions) issued by various authorized staff to implement day to day construction supervision within the scope of work; to confirm verbal instructions; and to record verbal job agreements. Site memos must not include directions to vary the scope of work or order a variation to the contract
- In an emergency situation where lives may be at stake and/or property damage may ensue if action is not taken immediately.

Drawings

The Contractor is required to perform all work in accordance with the contract which includes the Specifications and "Approved for Construction" Drawings. Where the "Approved for Construction" Drawings are revised or changed in any way, the contract may have been varied.

Such change may attract a change in price. All drawings issued to the Contractor must be reviewed for possible changes or variations to the contract and any such changes brought to the attention of the Engineer.

Schedules and Programmes

The Contractor must submit a construction programmed for the work and there is a requirement for periodic updating and revision. The programmed is to be submitted to the Engineer for approval within a specified time frame (refer to the relevant clause in the Conditions of Contract).

The "approval" must not be regarded as a casual matter. It is important that the programmed be prepared in detail, that it be thoroughly checked and that it be revised and resubmitted by the Contractor if it is not in accordance with the contract requirements.

The construction programmed must comply with the requirements of the contract for completing the work within the contract completion period.

Measurement of Work

The Engineer, working through site staff, is responsible for measurement and assessment of the work at regular intervals. (Refer to Procedure CA18 - Measurement and Certification of Quantities).

Control of Contractor's Sub-contracts

If the Contractor employs sub-contractors to carry out part of the work, all instructions must be directed through the Contractor.

It must be noted that neither the Engineer and delegated staff nor the Employer have any direct contractual relationship with the Contractor's sub-contractors.

Reports and Records

All site staff supervising the Contractor's work are required to report regularly on the work inspected.

Apart from keeping a personal daily diary, all conversations, meetings and agreements with the Contractor's staff should be recorded. A system of daily reports will be prepared. (Refer to Procedure CA11 - Daily Diaries and Reports).

The Construction Quality Control Manual includes checklists for each Specification Group, and these checklists will form an integral part of the Quality records for the project. (Refer to Procedure CA12 - Quality Control)

2.4 File System

A properly structured filing system is necessary to:
- Keep track of contractual correspondence and site records;
  - Avoid duplication;
  - Be able to easily locate correspondence and records;
  - Store all correspondence and records in an orderly manner.

  Standardized file numbers will assist in familiarity with the system.

File Reference Number

The File Reference Number consists of three groups:

Project reference number - (File No) - (Sequence No)

The first group, 5015001, is the project reference number and is standard for all correspondence on this project.

The second group, the File Number, is described below

1. Client
2. Contractor
3. Associates
4. Personnel
5. Financial
6. External

   The third group is the Sequence No. All correspondence is numbered from 1 onwards.

> **Example 1**: File Reference No
>
> The correspondence is addressed to the Contractor, and is the 0th letter from the Consultant
>
> **Example 2**: File Reference No
>
> The correspondence is addressed to the Client and is the 0th letter from the Consultant.

Procedures

1. All project correspondence will be filed on a regular basis, in the correct file and in proper sequence. All inwards correspondence will be received by the Office Manager and will be stamped with the date received and the distribution list. The relevant details will be recorded in the appropriate Register and the Inwards Sequence No from the register will be marked on the correspondence.
2. Both inwards and outwards correspondence will be placed on the one file in chronological sequence.
3. Once filed, no correspondence will be removed from a file unless for transfer to another file. In this event the removed correspondence will be replaced by a sheet denoting Letter Number(s) Transferred to File No. xxxx from dated…
4. Files will be kept in a lockable filing cabinet and a record of removal and replacement maintained.
5. A set of backup files will be stored in soft copy.

2.5 Contract Registers

Registers shall be created and maintained as electronic spread sheets for ease of updating and reproduction.

The following registers are to be created and maintained in the project office with a monthly print out:

Incoming Correspondence

Outgoing Correspondence

Contractual Claims

Variations

Issue of Drawings

Site Instructions

Responsibility: Administrator is responsible for the operation of all registers.

2.6 Contractual Correspondence

It is essential that all project staff be aware of their limits of authority in communicating with the parties to a Contract, of the significance of their written communication and of the formal or contractual communication required.

Contractual correspondence within the authority of the TL will be issued under signature of "TL", or if required under the Conditions of Contract, "The Engineer". The signature block on outwards correspondence is to read:

Yours faithfully,

Team Leader

All external correspondence must be signed by the TL and not by anyone else. If the TL is to be absent from site for a lengthy period, it may be necessary to revoke his authority and appoint an alternate during that period.

All internal correspondence should be in the form of Memos and not letters. The format to be used is given below.

The File Numbers are given in Section CA04 - Filing System

Procedures

1. Correspondence will be received at the project office.
2. All contractual correspondence received from a Contractor is to be

answered as soon as possible. If the reply will take longer than three days, an acknowledgement is to be forwarded immediately.

Responsibilities
TEAM will:
- Allocate file numbers and complete the distribution list for all correspondence;
- Sign all outwards correspondence; TL.
- Assign specialists for comment/review/action (eg Bridge Engineer, Contracts Specialist, etc.) or others as required.

TEAM will:
- Receive all inwards correspondence in English
- Stamp correspondence with the Inwards Sequence Number;
- Complete and maintain the Registers;
- Advise the Outwards Sequence Numbers for outwards correspondence;
- Copy and circulate correspondence;
- Maintain the files, and keep them secure.

2.7 Contract Commencement

Proper administration requires that certain matters be attended to at an early stage in the project.

Familiarisation

One of the first tasks for site staff when they arrive at the contract site is to become familiar with the site, drawings, specifications and the administration and inspection procedures to be adopted.

The TL will personally brief senior staff, covering such matters as:
- Line of authority;
- Duties and responsibilities of each team member;
- Outline of Contractor's organisation and appropriate levels of contact;
- Requirements with respect to correspondence, reports and records.

The DTL will further elaborate the responsibilities and duties of local staff on their arrival at site.

Contract Documents

The formal contract documents are the basis for the administration of each contract and only the specific requirements of those documents can be used in establishing the rights and obligations of both parties to the contract. All personnel involved in administration of the contract must ensure that both the Employer and the

Contractor receive fair and equitable treatment.

All other documents issued for the purposes of tendering, all interviews, negotiations and correspondence exchanged by the parties prior to the execution of the Agreement and not included in the contract documents have no contractual significance and are not to be used or quoted when interpreting the contract.

The TL will retain one set of contract documents including "Approved for Construction" drawings as a master set in safekeeping in the project office. This set of documents and drawings will be kept intact and updated with all issues of variation orders and drawing revisions so that a complete and up-to-date set of documents is available during the contract period (see Procedure CA30 - Work-as-Executed Drawings). Copies of relevant parts of the contract documents and drawings will be issued to site staff as necessary to perform their duties.

Pre-Start Meeting

A Pre-Start Meeting with the Contractor will be held in the early stages of the contract. This meeting will be chaired by the TL, and attended by the DTL, Contracts Specialist, and other specialists if mobilised.

The TL will prepare and distribute an agenda for the meeting, which will include the following:
- Responsibilities of the Consultant's staff
- Exchange of organization charts

- Contractor's Mobilisation Plan
- Submission of the Programme of Work from the Contractor
- Submission of proposed Cash Flow from the Contractor
- Submission of proof of insurance from the Contractor (refer Conditions of Contract)
- Security
- Contractor's Manning Schedule
- List of any sub-contractors, with corresponding applications for approval
- Contractual obligations of both parties
  - addresses for correspondence, and handling of same
  - issue and control of drawings
- Procedures for processing IPCs, measurement of quantities, agreement on indices
- Technical matters
  - setting out the works
  - inspections & testing of work and use of RFIs
  - construction procedures, including submission of Methodology Statements
  - issue of Site Instructions
  - submission of shop drawings if required
- Resources – their sources and availability (cement, reinforcing steel, bitumen, including test certificates)
- Requirement for joint survey
- Consultant staff facilities – accommodation, vehicles, communications etc.

• Frequency and location of Site Meetings

The TL will ensure that the procedures and agreements reached are within the terms and conditions of the Contract.

Minutes of this meeting setting out basic agreements and procedures will be recorded by the TL and distributed to all persons attending.

Emergency Telephone Numbers

The Office Manager will prepare a list of contact numbers of all senior personnel involved with the contract. A copy of this will be provided to the Employer, the Contractor and all site personnel and a copy will be posted in the site office in a prominent location.

Hand Over of Facilities to the Contractor

The Conditions of Contract and the Specification detail the facilities that the Employer will make available to the Contractor. Written and/or photographic records are essential to avoid conflict at the end of a contract in determining responsibility for clean-up and restoration of quarry areas, borrow pits, roads and on-site areas. These records will be prepared by the DTL prior to any site disturbance. Copies of records will be retained for ensuring that the Contractor meets his obligations on completion of the work.

Possession of Site

It is the Employer who gives possession of site to the Contractor. The Contractor is then responsible for maintenance of the existing facility during the construction period.

Climatic Conditions

Exceptionally adverse climatic conditions can generate claims for extensions of time and, in cases of heavy precipitation, flood damage which affects insurance claims may occur. Consequently, it is important to be aware of the effects of weather on a contractor's progress in such activities as earthworks, concreting and painting; its effect on labour attendance and stand downs, and damage to the works, the plant and equipment and temporary works.

Personal recollections of weather patterns are of no use for reporting or in assessing claims for extensions of time and only formally measured and recorded weather data are to be used.

The DTL will arrange for the collection and maintenance of daily weather reports. The provision and installation of weather recording equipment is generally included as part of the laboratory equipment to be provided by the Contractor.

The DTL will also obtain from the relevant authorities wherever possible historical records of rainfall and other weather details that relate to the contract site for at least the last ten years and longer if available.

EMERGENCY TELEPHONE NUMBERS
SENIOR PROJECT PERSONNEL

	Title	Work	Home	Mobile
Client				
	Manager			
Consultant				
	TL			
	DTL			
Contractor				
	PM			

Fire   _____

Ambulance   _____

Doctor   _____

Hospital   _____

Police   _____

Electricity Company   _____

Water Company   _____

Telephone Company   _____

Insurance Company   _____

2.8 Issue of Drawings and Other Document

The drawings depicting the scope of work are legal documents which form an integral part of the contract and the Contractor is required to perform the work in accordance with them.

Consequently, to achieve proper administration of the contract, the transmittal of drawings and other documents must be handled with circumspection and properly recorded at all stages to avoid unnecessary disputes and claims.

Procedures for Drawings

1. As part of the contract documents, the Contractor will be issued drawings which accurately reflect the work upon which the tender was based and accepted. This will avoid any claim or dispute that the drawings have been changed, revised or unseen by the Contractor before the contract was signed.
2. The Contractor will be issued with the number of sets of construction drawings as required by the Conditions of Contract. These drawings will be marked "Approved for Construction".
3. All drawings will be accompanied by a Drawing Transmittal form which is to be signed by the Contractor acknowledging receipt of the drawings.

When any drawing is revised, copies of the revised drawings must be issued to the Contractor with the revisions clearly marked, under cover of a Drawing Transmittal form.

Prior to issue, all revised or new drawings will be checked for variations from the scope of work or departures from the Specifications. The results of this check will initiate, if necessary, the appropriate procedure for issue of a Variation Order (Refer to Procedure CA26 - Variations).

If necessary, the TL will liaise with the Employer to ensure the Employer is aware of any drawing that is inconsistent with contractual obligations or that may give rise to a variation. This will give the Employer the opportunity to reconsider or confirm the revised drawing prior to issue to the Contractor.

After issue of drawings to the Contractor, the TL will ensure that the master set of drawings and Drawings Issued Register is updated (Refer to Procedure CA06 - Contract Registers).

Attachments: Sample Document Transmittal form Document Transmittal.

DRAWING TRANSMITTAL FORM

Contract Number: Contract Name:		Date of Issue									
		Day									
		Month									
		Year									
Drawing Title	Drawing No	Revision No									
No of Drawings Issued to:											
Legend		Client									
T Transparency		Contractor									
P Print		Site Office									
R Reduced size print		Other									
Purpose of Issue:		Information									
		Approval									
		Comment									
		Construction									
		Quotation									
		Other									

Attached drawings have been checked and agree with above list:

Issued by: _____

Receipt Acknowledged: _____ Date: _____

DOCUMENT TRANSMITTAL

To: _____   Date: _____
    _____   Your Ref: _____
    _____   Our Ref: _____
Attention: Project Manager   Contact: _____
Project: _____   Phone: _____

Herewith: _____
_____
_____
_____
_____
_____
_____
_____
_____
_____
_____
_____
_____
_____

For: ◦ Information   ◦ Resubmit   ◦ Return to you
        ◦ Approval   ◦ Construction   ◦ Other
        ◦ Comment   ◦ Quotation

Attachments documents have been checked and agree with above list:

Issued by:

_____ : Date _____

Receipt Acknowledged:

_____   Date: _____

2.9 Subcontractors

The Conditions of Contract preclude subcontracting without the consent of the Employer.

Procedures

If a Contractor advises that he proposes to use a subcontractor, the following details are to be obtained and submitted to the TL, who will then assess these factors before a submission is made to the Employer as to the suitability of the subcontractor:
- Ability to do the work.
- Previous experience
- How much of the work is being subcontracted, i.e. the more work being sub-contracted the more information that will be required from the subcontractor.
- Work Methods
- Staffing
- Safety record
- Insurances
- Financial capacity.

The TL will prepare the submission to the Employer.

In the event of the subcontractor working on site but not being approved, the TL will notify the Contractor and the Employer of the breach of contract.

Responsibilities

Site supervision staff are responsible for checking whether subcontractors have been approved and advising the TL of any non-approved subcontractors working on the site.

The TL is responsible for following up with the Contractor to ensure that the provisions of the Contract are followed, and for advising the Employer.

2.10 Site Instructions

For minor matters, or when insufficient time is available to issue a formal letter, authorised staff may issue a Site Instruction to the Contractor.

Procedures

Staff who have the necessary authority may issue instructions to the Contractor.

Instructions may only be issued for work that is specified in the Contract and not for any new work or any activity that could be construed as a variation or result in an extension of time claim.

Instructions must refer to the relevant Specification clause or drawing.

A copy is to be filed in the dedicated site file and the Site Instructions Register completed.

Responsibilities

Authorised staff are responsible for completing the Instructions, maintaining registers and ensuring correct distribution of copies.

Attachments: Site Instruction form.

2.11 Daily Diaries and Report

Consultant site staff are to maintain two types of records:
- Daily Diaries
- Daily Reports

Daily Reports are to be recorded on a Consultant's form and are used to maintain a record of significant events and activities for historical purposes that later may be used in preparing more detailed reports or addressing contractual claims.

All professional contract supervision staff must also maintain a daily diary in which notes and records of daily activities and conversations are kept. The diary is not a substitute for the Daily Report and needs to record:

- General daily activities. Every day must be recorded. If there is no work carried out on a given day, or the recorder is absent for any reason, these details need to be recorded;
- Telephone calls made or received with details of the conversation;
- Details of all substantial conversations held with the Contractor, as well as any instructions issued and commitments made by either party;
- Any work or material not conforming to the specified requirements, as well as the action taken;
- Unforeseen conditions or other problems that may affect the Contractor

Procedures

As well as maintaining a comprehensive diary, professional supervision staff will compile a Daily Report recording daily events on the aspects of the work for which they are responsible. The report will be prepared for all normal working days irrespective of whether or not any progress was accomplished.

The following information will be recorded:
- Areas/structures where work is in progress, with details of work being performed and the major plant and approximate number and classification of men involved;
- When work commences in a new area or on a new structure;
- When work has stopped for a period, and reasons for stoppage;
- Concrete placements achieved, including start and finish times;
- Subcontractors on site and location of areas where working;
- Details of any lost time due to industrial disputes, weather etc;
- Arrival/departure of major items of plant;
- Significant non-availability of major/critical items of plant;
- Arrival of major items of material;
- directions/instructions given to the Contractor, other than those of a routine nature;
- Any significant changes to the Contractor's supervisory staff.

Site staff will forward their reports to the DTL on a daily basis. The DTL will note any matter that requires follow-up action, and advise the TL.

The Office Manager will maintain files of the Daily Reports.

Daily information such as rainfall, site delays and visitors will be recorded by the DTL on a Miscellaneous Information form. These forms will be retained and filed with the Daily Reports.

Responsibilities

All senior staff are responsible to maintain a daily diary.

Professional supervision staff are responsible for preparing Daily Reports on all areas of the work for which they are responsible.

The DTL is responsible for reviewing the Daily Reports on a regular basis and taking action where necessary and for completing a Miscellaneous Information form.

The Office Manager is responsible for filing the Daily Reports and the Miscellaneous Information forms.

Attachments: Miscellaneous Information Form Daily Dairy

2.12 Quality Control Field Inspection

A major part of the supervision role is to ensure that construction is carried out in accordance with the specified standards.

Procedures

Procedures for field inspection have been developed by the TL in accordance with the requirements of the Specifications. These procedures are given in the separate Quality Control Manual.

Procedures will include check lists to permit the systematic checking, inspection and acceptance of works, record sheets to record the details of an activity as it is happening and report sheets to summarise the results of an activity after its completion.

The date and findings of all inspections will be recorded, together with pertinent details of remedial work or corrective work directed. Follow-up inspections will be carried out to verify that the remedial work has been carried out, and findings again recorded. The check lists and record sheets are to be completed as the work is in progress. Inspection staff must not write information on other pieces of paper or notebooks and later transfer them to these sheets, rather they must be completed at the time the event is occurring and, whilst care must be taken to keep the sheets neat and legible, it is understood that in the course of the work they may become soiled or creased.

The Contractor will be advised immediately, by the issue of a Non-Conformance Notice, if any inspection reveals that work does not conform to the requirements of the Specification.

The TL will ensure that the Contractor corrects all non-conformances. After the approved rectification work is completed, the Non-Conformance Notice is closed out by completing the form and the register.

Responsibilities
The TL is responsible for:
- Ensuring that site staff are inspecting the works to the appropriate standard;
- Ensuring that all tests, inspections etc. are recorded on the appropriate forms;
- Ensuring that Non-Conformance Notices are properly issued and finalised.
- Ensuring all work is inspected as and when necessary and that all work conforms with the requirements of the Specification;
- Ensuring all inspections are recorded on suitable forms;
- If authorised, signing Non-Conformance Notices;
- Ensuring rectification work is properly performed to finalise Non-Conformances;
- Setting up and maintaining a Non-Conformance Register.

All field staff are responsible for quality control checking and identifying any Non-Conformance.

Attac: Non-Conformance Notice Non-Conformance Register

The Contractor is hereby notified that after inspection by the Engineers Representative that:

☐ Traffic Management Safety
☐ Material Tests
☐ Inspection
☐ Survey Check
☐ OTHER

Engineer confirms that the Works do not conform to the Contract Specification/Design Drawings/Contract Provision.

The Contractor is required to rectify and confirm the proposed remedy in writing to the Engineer with 24 hours of this notice.

NONCONFORMITY REPORT

QUALITY DEPARTMENT SECTION:	*Highway section*	REPORT No.:	NCR
CONTRACTOR / SUBCONTRACTOR:	CSCEC LTD. Georgia Branch		
REPORTED BY:	SIGNATURE:		
LOCATION:	DATE:		

1. NONCONFORMITY DESCRIPTION	(Name and position of higher responsible present on site) (Attach witness list if any) (Detailed description of the nonconformity) (Attach persons involved) (Attach a sketch of the nonconformity)
At the exit from tunnel No. 4, the asphalt concrete pavement level does not correspond to the concrete base level, which could create a hazard for traffic.	
2. PROPOSED CORRECTION	(Detailed description of the corrective action) (Attach static calculations if necessary) (Attach new construction drawings if needed) (Attach materials technical certification if needed) (Attach any other document for Engineer information)
The contractor should cut the concrete at the asphalt level.	
FULL NAME:	Signature:

IS IT POSSIBLE TO SOLVE THE NONCONFORMITY IMMEDIATELY ON-SITE?	YES ☐	NO ☐	IF **YES**, MARK NONCONFORMITY LEVEL	1 ☐
IF **NO**, MARK NONCONFORMITY LEVEL	2 ☐	3 ☐	4 ☐	AND THE FOLLOWING LINES MUST BE COMPLETED;

3. INTERNAL APPROVAL AND COMMENTS OF THE CORRECTIVE ACTION PROPOSED

___
___
___
___
___
___
___

FULL NAME: _____ *CONSTRUCTION MANAGER / PROJECT MANAGER* Signature: _____

4. APPROVAL BY THE ENGINEER OF THE CORRECTIVE ACTION PROPOSED

___
___
___
___
___
___
___

FULL NAME: _____ Signature: _____

5. CONTROL OF THE APPLICATION OF THE CORRECTIVE ACTION BY THE ENGINEER

☐ APPROVED
☐ NOT APPROVED

REMARKS:

FULL NAME: _____ Signature: _____

6. CLOSURE OF THE NONCONFORMITY AND TRANSMISSION TO:

QUALITY DEPARTMENT OF THE CONTRACTOR _____
signature and date received

ENGINEER _____
signature and date received

NON CONFORMANCE REGISTER

Non Conformance Register

No	Date	Non Conformance Particulars	Method of Rectification	Date Approved

2.13 Survey Checks

Survey control is the prerequisite for quality in contract work. The Contractor's work has to be checked to ensure that it is to the dimensions and tolerances required by the Contract, even though the Contractor is fully responsible for the accuracy of setting out and his work generally.

There may be certain stages of construction where the Contract specifically requires the Contractor to stop and await a survey check and approval/acceptance before covering up the work. If the Contract does not specifically have such a requirement, the TL can direct the Contractor not to proceed beyond a certain construction stage (and cover up work) until a survey check has been done, and approval/acceptance has been given.

Procedures

The TL will review the Contract Documents to identify any mandatory check points requiring approval/acceptance and will compile a list of requirements.

For all road work, a joint survey is to be undertaken prior to the commencement of earthworks, and signed off as agreed by both Consultant and Contractor surveyors. Thereafter, all survey for measurement of work will be done jointly, agreed and signed off.

Standard procedures will be agreed between the Consultant and the Contractor.

If survey checks show that the work does conform to the specified requirements, the Surveyor will advise the Team Leader or DTL and the Contractor and note this in his survey record. This will be followed by confirmation in writing.

If results are non-conforming, the TL will decide either:
for a minor Non-Conformance, the Contractor may either leave as is or carry out some minor re-work, or

- For a serious Non-Conformance, re-work, etc., and resubmission is necessary.

All Non-Conformances will be advised in writing to the Contractor using the Non-Conformance Notice (see Procedure CA11 - Quality Control - Field Inspection).

Responsibilities

The TL is responsible for ensuring that any check or hold points reflect the requirements of the Contract and that the system is followed. The TL is responsible to inform the Contractor of these check or hold points.

The TL and the DTL are responsible for:
ensuring that the system works on a day-by-day basis;
- ensuring survey check results are submitted promptly to the Contractor; and

- ensuring that all areas/sections that are to be checked, do in fact get checked and eventually conform to requirements, and that any Non-Conformance is correctly finalised.

The Surveyor is responsible for notifying the TL or the DTL and the Contractor of any Non-Conformance in survey work.

2.14 Site Meeting and Minutes

Regular, properly conducted meetings, with agreed minutes, ideally will provide the basis for good cooperation between the Consultant and the Contractor, and keep correspondence between the parties to a minimum.

Procedures

Site meetings will be held on a regular basis at two levels, to be attended by the TL, the Contractor's Project Manager, an Employer's representative, and any others that may be deemed necessary to resolve particular problems:

   a. Weekly Site Meetings to discuss technical issues; and

   b. Monthly Progress Meetings.

The frequency, attendees etc., to be adopted for meetings will be agreed at the Pre-Start Conference to be convened soon after the award of the Contract (see Procedure No CA07 - Project Commencement).

Meetings will follow a standard agenda including the following general topics:

- Matters arising from previous meeting and confirmation of previous minutes;
- Review of progress and any problems;
- Review of the Programme of Work;
- Outstanding correspondence;
- New matters raised by Contractor;

- New matters raised by the Employer;
- Next meeting.

Discussions on individual matters should not be allowed to become too prolonged. If an issue cannot be resolved satisfactorily or requires further discussion, a special meeting should be arranged to deal with the topic. This will then allow the meeting to progress onto the other matters on the agenda.

The regular Site Meeting should not be used for discussing claims or contentious issues, these being left to special meetings, possibly with fewer people in attendance, and after further research.

As soon as possible after the meeting a set of minutes will be prepared by the TL. Minutes will be limited to the recording of progress, agreements, approvals, requests and decisions. The minutes will be forwarded to the Contractor and the Employer's representatives.

If the Contractor or the Employer's Representative do not agree with any point in the minutes, formal notification should be made as quickly as possible such that it is discussed at the next meeting and resolved.

Responsibilities

The TL is responsible for:

- Convening the meetings and preparing and circulating an agenda;
- Chairing the meeting;
- Preparing and issuing the minutes.

The DTL and senior Consultant engineers are responsible for advising the TL of any items needed to be discussed at the meetings.

Attachments: Sample of Agenda Minutes of Meetings Proforma

MONTHLY PROGRESS MEETING NO.
MEETING AGENDA

Agenda for Meeting No.   to be held in the Office at   . hours on   .

1. Introduction
2. Matters arising from Previous Meeting
3. Progress of Works
    a. Mobilisation
    b. Temporary Works
    c. Permanent Works: [Separate into main features]
4. Programme of Work - Review
5. Measurement and Interim Payment Certificates
6. Issues, Claims, Variations and Extensions of Time
7. Reporting
8. Outstanding Correspondence
9. New Matters raised by Contractor (Technical and Contractual)
10. New Matters raised by Consultant (Technical and Contractual)
11. New Matters raised by Employer (Technical and Contractual)
12. Date of Next Meeting

2.15 Monitoring of Progress

Regular monitoring and documentation of the progress of construction is essential to provide a measure of contract performance, provide a record of progress and be useful for an analysis of delay claims.

The Conditions of Contract require the Contractor to submit, prior to commencement, a Programme of Work for the approval of the Engineer. This programme is also subject to regular updating by the Contractor.

It is essential that the critical path for the contract work be identified. An update of the programme may result in change to the critical path, and the reasons must be recorded so that the information is available for any dispute that may arise.

There are two methods of measuring progress:
- Physical
- Financial

Both physical and financial progress percentages are to be included in the monthly reports.

Procedures

At the start of the contract, the Contractor has to submit his

Programme of Work.

The Contractual Programme will be fixed as the "baseline" programme and actual progress input will be made periodically to give a comparison between planned and actual progress. The progress data will be entered at least monthly but shorter periods may be preferable if the activities are of short duration and critical.

As they occur, delays will be input and analysed to determine whether or not an extension of time is applicable.

The DTL, or his representative, will maintain a special set of drawings and/or schedules on which progress, date of placement, erection or installation, date of inspection and of acceptance of various features of the work will be recorded.

Drawings and/or charts will be prepared for each major item of work such as, but not limited to, the following examples:
- Excavation
- Culverting
- Earthworks
- Road pavements
- Bridge substructures
- Bridge superstructures

Physical progress will be assessed as follows:
- Each major item of work will be given a percentage based on its BOQ value compared to the contract price [eg value of AC/total contract value];

- An assessment of completed, acceptable work will be made for that item and applied to the percentage above

Financial progress will be assessed as the value of IPCs submitted as a percentage of the total contract price.

Responsibilities

The TL and DTL are responsible for setting up and monitoring the overall project progress, and preparing progress details for the Project Monthly Reports.

The TL or his nominated representative is responsible for monitoring progress, assessing physical and financial percentages complete, and maintaining detailed records.

2.16 Progress Report

The Consultant is required to produce a
number of reports as part of the services to the Employer.

Monthly Reports

A Monthly Report has to be prepared by the TL from a project perspective. The Monthly Reports are the means whereby the Employer and the ADB are kept formally conversant about the project and individual contracts on a regular basis. The Monthly Report will include:

- A summary of the progress of the construction contract for the reporting period, including percentages complete in both physical and financial aspects, and will note specific events during the period;
- A summary of the Consultant's activities during the month, including personnel, financial aspects and any outstanding issues.

Senior Consultant engineers will prepare details within their respective fields and submit to the TL by the third day of the following month [eg, for the May report, details are to be submitted by 3 June].

Construction contractual details required include:
- Physical Progress: Construction progress during the month, and cumulative to date, drawing specific attention to any major causes of delay (administrative, technical or financial) with details of remedial action taken or recommended;

- Financial: A comparison of actual and forecast expenditure both during the month and cumulative to date; a record of the status of payment of the Contractor's monthly invoices; all claims for cost or time extensions;
- 'S' Curve: The 'S' Curve is to be updated each month;
- Quality: Summary of all tests conducted during the month. Technical appreciation of any design or quality control problems including details of remedial action taken or recommended;
- Ancillary Actions: Status of compliance with the Environmental Management Plan and Resettlement Action Plan, and updates of the Environmental Monitoring Checklist and the LARP Monitoring Checklist;
- Any Major Issues: Details of any problem areas that have developed - these could be ongoing from previous months.
- Photographs: A maximum of four progress photographs.

Other Reports

In addition to the Monthly Reports the following reports are required to be produced:
- Design Review Report
- Annual Report
- Interim Contract Completion Report
- Project Completion Report

Both the Interim and Final Completion Reports will include the following details:

- Implementation - construction equipment and plant, personnel, history of contract;
- Financial details - summary of IPCs [date of claim, corresponding payment date, amounts, etc.]
- Defects list at time of Taking Over;
- Summary of claims - dates, reason, resolution
- Recommendations

The TL is responsible for the final preparation and submission of all reports.

Responsibilities

The TL is responsible for the preparation and submission of all reports required under the Consultant Contract.

For the Monthly Reports, senior Consultant engineers will prepare details within their respective fields and submit to the TL by the third day of the following month [eg, for the May report, details are to be submitted by 3 June].

Attachments: Sample Monthly Contract Report - Table of Contents
Sample Monthly Contract Brief

2.17 Financial Progress Monitoring

Effective contract management requires a system that will provide a realistic forecast of expenditure.

The Conditions of Contract require the Contractor to submit prior to commencement of the Works, a Cash Flow Estimate for the information of the Engineer and the Employer. This Cash Flow Estimate is also subject to regular updating by the Contractor.

The Cash Flow Estimate is generally submitted and revised with the Works Programme, and when using project management software such as Primavera, it is a key component that is generated as part of the Programme.

Procedures

The TL and DTL will check the theoretical cash-flow estimates submitted by the Contractor based on the approved Programme of Work and major BOQ items.

Actual expenditure based on Interim Payments will then be plotted over the duration of the contract and compared with the theoretical cash-flow estimates.

When actual expenditure varies by 5% or more from the predicted cash-flow, the TL may request the Contractor to submit a revised cash-flow estimate.

Responsibilities

The TL and DTL are responsible for ensuring that the Contractor submits cash-flow estimates, for checking the accuracy of these cash-flow estimates and for monitoring actual expenditure.

2.18 Measurement and Certification of Quantities

In accordance with Clause 12.3 of the Conditions of Contract, the Engineer is required to determine the value of the Works.

The intention of this procedure is to ensure that correct IPCs are produced quickly and that a minimum of effort is required to finalise quantities at the completion of the Contract.

Procedures

The Consultant's and Contractor's staff must meet regularly to review measurement records and quantity calculations. These meetings must ensure that all items are measured and measurements, calculations, etc., are correct. Items for which agreement cannot be reached will be recorded. Payment will not be made of the quantity calculated by the Consultant's staff until the differences are resolved.

The system of measurement adopted for each item will be agreed between the TL and the Contractor's Project Manager. These systems will be chosen for practicality and economy in use of staff and accuracy in meeting provisions of the Contract.

Quantities upon which payments will be made fall into three categories:
- Quantities calculated from drawings using measured field dimensions and levels
- Quantities physically measured in the field
- Quantities based on weight either calculated or physically weighed.

Calculated Quantities

Quantities for such items as concrete paid for on the basis of volume will be calculated from pay lines shown on the Drawings, within the limits specified for the appropriate items. All calculations will refer to the appropriate drawings used and to field directions, if issued, to amend or clarify drawn dimensions. Field measurements will indicate work under each item completed within the specified limits.

Physically Measured Quantities. In all cases the measurement should be witnessed by a representative of the Contractor and jointly certified.

Quantities Based on Weight

Where the Contract calls for items to be paid at a unit rate per kilogram or tonne, the methods of measuring such items will be agreed with the Contractor. The following guidelines will be used:

computations of mass will be used only for recurrent items of regular shape which cannot be easily weighed, i.e., mass of structural steel shall be based on dimensions shown on shop drawings and the mass calculated using 7,850 kg/m3:

individual small items will be weighed, if convenient, on approved scales which can be checked;

the mass of large items of equipment or machinery will be accepted on the basis of weighbridge dockets which must record the mass of all tare blocking and packing. Alternatively, manufacturer's shipping weights will be acceptable;

the mass of multiple items such as reinforcing bars, rails,

pipe specials and valves, etc., will be accepted if obtained from manufacturer's lists or catalogues. Very small items such as nuts, bolts, washers will be weighed in bulk and averaged to obtain individual mass if necessary;

the mass of coating materials, paint, gaskets, welding runs, grout and caulking materials applied at site are not included in computing mass for payment. Waste, off cuts and rolling margins will be disregarded.

Specific Methods of Measurement

The following are specific methods which will be used for certain items of work to obtain the necessary accuracy and avoid duplication of payment. These methods relate mostly to items which cannot be checked later and require formal check out when the work is performed.

Reinforcing Steel

Reinforcement will be checked in place prior to placing concrete. Measurement of the bars may be computed from bar bending schedules in advance but changes made in the placement will be recorded. Consequently, the Contractor's claim for payment will be accompanied by calculations for the original placement and a certified copy of the "Adjustment to Reinforcing Bars" Form if necessary.

Embedded and non-embedded materials

Where a concrete placement includes the embedment of metal work, water-stops, ducts, brackets, etc., the Contractor is responsible for ensuring correct installation in the placement. Site staff are required to verify this before placement. Measurement will be made at the time of this verification using the "Embedded Materials Check-out Sheet". Only the materials actually checked in each placement will be recorded, i.e. where a pipe traverses more than one placement only the actual length in the placement will be recorded, otherwise duplication may occur. The form will be filled out in duplicate and jointly signed by both parties. The remaining details will be completed prior to submission of a claim for payment and will be accompanied by the relevant forms for each item claimed. The form will be prepared in

advance as a check list prior to inspection, listing every item to be installed.

Surveyed cross-sections

Where excavation is made to lines not actually defined on the Drawings or where large mass excavations are to be paid on the basis of surveyed cross sections, or for embankments, the procedures for measurement will be as follows:

a survey will be made of the original ground surface (after stripping) and before excavation or filling is commenced. Further surveys will be made when portions of the work are to be measured for payment;

all surveys will be made jointly with the Contractor;

Consultant and Contractor surveyors will prepare the appropriate plotted outputs, compare and agree them; contract pay lines will be superimposed where required and the pay quantities computed to obtain agreed quantities. The pay line for embankments will be the design batter line, even though the Contractor must "overfill" to ensure that he attains compaction out to the design batter line;

all survey notes, computations and plotted outputs will be signed by the Consultant's and the Contractor's surveyors.

Quantity Survey Form

This form is the basis of payment and is a contractual document which will be treated as a record for safe keeping until the Contract is finalised.

The following will be recorded:
- BOQ item being measured;
- The location of the work performed;
- The measurements recorded;
- The unit of measurement for payment; and
- The agreed quantity.

No alteration is to be made to a QS form unless both parties sign the alteration.

All QS forms will be numbered to identify the BOQ Item and the number of QS forms issued for that Item as follows:

QS forms will be filed under individual Bills, e.g. all QS forms issued for Bill 5 will be filed together, those for Bill 6 in a separate file, etc. Subdivision into separate items will be done by using dividers. The front of each file subdivision will contain a summary of quantities paid under that item identified by monthly payment and accumulated total.

Pay quantities in Interim Statements will be rounded off to the nearest unit. The final Progress Payment will be certified to the nearest tenth of a unit.

Preparation of the QS Form

The manner in which a QS form is made out cannot be standardised but the following basic rules will apply:

both the Contractor and site staff will sign any alterations made to QS forms;

cancellation to only part of a form will not be made but the whole form will be cancelled or superseded. A new QS form for the same work will record that a previously issued form has been cancelled;

where the work is completed the QS form representing the final measurement will be recorded as "FINAL" and "Work Completed" written on the form;

if an actual measurement is not made and a provisional payment is agreed upon with the Contractor the QS form will state "Provisional Payment" and the agreed quantity to be paid. All previous provisional payments will be cancelled by the issue of a QS form when measurements and/or calculations are made to finalise a pay quantity;

- Supporting data, cross sections, drawing numbers, field directions, etc. will be listed for identification in support of the QS form.

Calculations

All calculations will be made on standard Calculation Sheets. It is essential that calculations are clear, concise and set out to make for easy verification and cross checking. The accuracy to which calculations are made will be determined firstly by the order in which inaccuracies are magnified and secondly by the unit rate at which the relevant quantity will be paid.

As a general rule the following will be adopted:
- Where constants or factors are adopted they will be agreed with the Contractor;
- Quantities will not be rounded off in calculations. This will be done by the TL as set out above.

Responsibilities

The TL is responsible for review of the Interim Statements and preparation of the IPC for each contract.

The senior Consultant engineer of each major facet of work (earthworks, pavement, structures, tunnels, etc.) is responsible for ensuring that the physical measurement on site of the Contractor's work is carried out. He will delegate specific responsibilities to site staff for the taking of field measurements and the preparation of calculations to support the quantities certified for payment. He will prepare specific procedures for implementing submission of such data and reconcile all differences with the Contractor on measurement and payment.

Surveyors are responsible for carrying out all survey measurements and calculations.

Attachments: QS Form
Summary of Quantities Form
Adjustment to Reinforcing Bar Quantity Form
Embedded Materials Check-out Sheet

QUANTITY SURVEY FORM - BOQ ITEM No. _____

QS Form No.   Cxxx/XXXX/YY Sheet ____ . of ____
Description of Work: _____
Location: _____
Relevant   Drawing Nos: _____

> Where applicable, sketches, measurements, calculations, etc to be provided in this space

_____   _____
Contractor   Consulting Engineer

Date: _____   Date: _____

SUMMARY OF QUANTITIES

Sheet ____ . of ____

BOQ Item No. _____

Quantity in BOQ: _____   Rate in BOQ: _____

Date	Location	Dwg No	QS Form No	Quantity	Sub Total	Cumulative Total

_____   _____

Contractor   Consulting Engineer

ADJUSTMENT TO REINFORCING BAR QUANTITY

Sheet _____ . of _____

Feature: _____ Refer QS Form No: _____

Placement Location: _____ Drg No: _____

ADDITIONS							
Bar Mk	Dia. (mm)	No Off	Additional Length /bar (m)	Total Additional Length (m)	Mass per m	Additional Mass (tonne)	Reasons for Change
TOTAL ADDITIONAL MASS							
DEDUCTIONS							
Bar Mk	Dia. (mm)	No Off	Additional Length /bar (m)	Total Additional Length (m)	Mass per m	Deducted Mass (tonne)	Reasons for Change
TOTAL DEDUCTED MASS							
TOTAL NETT ADDITION/(DEDUCTION)							

_____

Contractor

_____

Consulting Engineer

EMBEDDED MATERIALS CHECK-OUT SHEET

Sheet ____. of ____

Drawing		Description	Unit	Checked by		BoQ Item No	
No	Item No			Office	Field		

Placement: _____   Date Placed: _____

_____   _____
Contractor   Consulting Engineer

2.19 Interim Payment

Payment can only be certified by the person with the delegated authority and that certification must be made within the time set out in the Contract. Checking of the interim Statements must be done in a timely manner within these constraints.

Procedures

Work will be measured for payment in accordance with Procedure CA18 - Measurement and Certification of Quantities.

The TL will discuss quantities and amounts with the Contractor before an Interim Statement showing the quantities and value of work done for the month is submitted by the Contractor.

The format of the Interim Statement will follow the format of the BOQ and should be prepared using an electronic spread sheet (Excel) for ease of transmission and review.

If there is no agreement with the Contractor on quantities or value, the TL will correct the Statement submitted by the Contractor and prepare a new Interim Statement with the corrected quantities and/or values.

The TL should also seek the concurrence of the Employer's Representative on site for the quantities and value of the work to avoid subsequent changes to the final statement. Obtaining this concurrence however must not delay the finalisation of the Statement so that the period stated in the Contract for submission of the IPC is exceeded.

When satisfied that the Statement is correct, the TL will certify and forward the Interim Statement to the Employer within the time limit specified in the contract.

Attachments: Sample Interim Payment Certificate

2.20 Variations

Variations occur where there is a change to the scope of work. They can therefore involve an increase or a decrease to the contract value. It is important that, once it is realised that a variation is required, the instructions must be issued promptly to minimise any adverse effects this may have on the overall works.

Variations to contracts in this project require the approval of the Employer. Because this process can be time-consuming, it is important that the Employer be presented with all known information and details about a variation so he is in a position to make a prompt decision.

Excessive variations can result in claims for payment over and above the cost of carrying out the work. Such claims may include increased overheads, extensions of time and other ripple effects not foreseen when individual variations are being processed.

Care must be taken with the timing of issue of variations. If the contract period has passed, the issue of a variation could affect the recovery of Liquidated Damages. The actual situation will depend on the particular contract conditions.

Procedures

If "Approved for Construction" drawings and/or specifications are revised, the TL will determine if the change constitutes a variation to the Contract, its extent and the terms and conditions under which the variation will be implemented. Extra work should not be ordered or changed drawings issued unless accompanied by a VO.

Approval of the Employer will be sought.

If possible, the value of the variation will be agreed with the Contractor before the variation is ordered.

If the Contract does not contain any applicable rates or prices for the varied work, the TL will forward a Request for Price form to the Contractor seeking new rates or prices.

If, after consultation with the Employer and the Contractor, a suitable rate or price cannot be agreed, the TL will determine a provisional rate or price for the work. This will be notified in writing to both the Contractor and the Employer.

Responsibilities

The TL is responsible for:
- Determining if a change constitutes a variation to the Contract, its extent and the terms and conditions under which the variation will be implemented;
- Obtaining the Employer's approval for the variation;
- Issuing a VO if holding the authority. If not holding the authority, forward the pertinent documentation to the appropriate person for signature;
- Determining the valuation of the variation.

The senior Engineers in charge of their facet of work are responsible for ensuring that work associated with a variation is completed.

Attachments:
Variation Checklist
Request for Price Form
Sample Variation Order

VARIATION CHECKLIST

Issue No:	
Type of variation:	
Location:	
Work described in Specification?	
Work shown on Drawings?	
Variation recommended?	
Applicable rates or prices in BOQ?	
Request for Price Form issued?	
Contractor's price proposal received?	
Will variation effect Date for Completion?	
Value of variation agreed with Contractor?	
Effect on time agreed with Contractor?	
Provisional rate necessary?	
Provisional rate determined?	
Rate fixed by Team Leader?	
Employer's approval necessary?	
Employer's approval received?	
Variation No:	

REQUEST FOR PRICE No……

In accordance with the provision of Clause …… of the Conditions of Contract, the Contractor is hereby requested to submit, for the Team Leader's consideration, a quotation for the item(s) of work listed in the schedule below:

1. SCHEDULE

Item No	Description	Unit	Quantity	Rate	Amount

2. DRAWINGS

State drawing numbers

3. SPECIFICATION

   Title

   Clause

   Requested by:

Team Leader   Date

SAMPLE VARIATION ORDER

[File No]

[Date]

[Contractor Address]

Dear Sir

In accordance with Clause ······ of the Conditions of Contract, you are instructed to do the following:

- [specify].

The work shall be carried out in accordance with the Specification ······. and with Drawing No _____

The value of this variation is agreed at _____. The valuation includes allowance for the following:

- All direct costs associated with the varied work;
- All indirect costs and profit associated with the varied work;
- All impact costs associated with the varied work; and
- All costs associated with any ripple effects resulting from the varied work.

No adjustment to the Date for Completion will be made by reasons of the changes and/or additions instructed herein.

OR (if no price agreed)

The effect if any of this variation shall be valued in accordance with Clause _____.

Yours faithfully

Team Leader

2.21 Environmental Factors

In accordance with the Contract documentation, the Contractor has to ensure that all activities are carried out in conformity with statutory and regulatory environmental requirements. The Contract Documents take precedence over this section and in case of any conflicts between the relevant clauses of the contract specifications and the environmental requirements outlined in this section, these should be referred to the Team Leader for clarification.

Procedures

The Environmental Specialist will produce the Environmental Monitoring Checklist.

The Contractor has to produce an Environmental & Social Management Plan (C-ESMP), which has to be approved by the TL before implementation.

Approval will be required for various activities to be undertaken by Contractor before the start of construction. These include identification of locations for siting stone crusher units, asphalt mixing plants, concrete batching plants, and borrow areas, quarry sites, workers camp, workshop and vehicle maintenance yard, material storage, waste disposal and disposal areas for removed asphalt pavement materials. Environmental considerations will have to be included in the locality and approval process, with particular respect to fuel and lubricants.

The following aspects require periodic inspection and supervision during the execution of works:
- Borrow area operation and management
- Quarry area operation and management
- Material storage area management
- Topsoil preservation and management
- Construction material re-use and disposal
- Workers camp management
- Soil erosion and sediment control
- Drainage control
- Air, noise and water pollution control and monitoring
- Occupational health and safety
- Preservation of cultural properties
- Environmental enhancement
- Disposal of old asphalt pavement materials.

At the end of each month, the Environmental Monitoring Checklist will have to be filled in and submitted as part of the Monthly Report A copy is attached.

Responsibilities

The Environmental Specialist will prepare the Environmental Monitoring Checklist.

The DTL is responsible for ensuring that the monitoring on site of the Contractor's compliance with his EMP is undertaken.

The TL is responsible to approve the Contractor's Environmental Management Plan.

ENVIRONMENTAL CHECK LIST

Item	Yes	No	Corrective Action
Earthworks			
Are vehicle movements restricted to designated routes?	✔		
Are speed limits for trucks established and followed?	✔		
Is surplus excavated temporary or permanent excavated material recorded in accordance with the submitted Contractors Excess Material Disposal Plan and are all excavated materials kept to designated temporary or permanent areas, levelled and drained?		✔	
Are soil erosion control methods in place?		✔	
Drainage			
Are worksites adequately drained to control run-off?		✔	
Are fuel/oil storage tanks contained in areas to control spillage?	✔		
Is there any visible evidence of surface water pollution?	✔		
Is wastewater from the work sites or camps properly disposed of?		✔	
Waste Management			
Are waste containers and waste materials properly stored, labelled and separated?		✔	
Are hazardous wastes properly contained?		✔	
Is the removal of waste/rubbish/garbage site wide and from various plant yards batching plant Contractors camp sites properly controlled and disposed of in compliance with the local legislation?		✔	
Water Pollution			
Has testing of existing river streams and creeks water been undertaken		✔	
Is any run off from the Contractors facilities degrading the water quality of any existing river streams and creeks water ways	✔		
Noise and Air Quality			
Is mechanical plant in good condition?	✔		
Are emissions from batching plants (concrete & asphalt) properly controlled?		✔	
Are haul roads/new roads watered to control dust?		✔	
Are maintenance areas kept free of dust?		✔	
Are procedures in place to keep noise levels down, particularly in urban areas?		✔	
Reinstatement			
Have natural contours and drainage channels been restored as far as possible?	✔		
Has all construction waste and rubbish been removed/buried?		✔	
Have borrow areas been regraded and smoothed?		✔	

2.22 Public and Relations; Health and Safety

Public Relations

Because of the construction activities taking place on the site, the general public and visitors should be made aware of possible risks that could face.

To alert the general public, the Contractor has to erect contract signs at either end of his site works, and specifically with road works, the Contractor must display warnings signs and/or flagmen in possible hazardous areas, as well as suppressing dust on unpaved roads.

Site visitors have to be accompanied by Consultant's staff member delegated for this task. Safety helmets and reflectorized vests should be provided where appropriate. Details of visitors are to be included in the Daily Diary.

Whilst not attempting to discourage members of the media, any requests for interviews or statements or comments must be directed to the Employer.

Health and Safety

Emergency situations may develop on site as a result of local unrest, warfare, flooding and earthquakes. An evacuation plan has to be prepared and all staff made aware of possible evacuation procedures.

Consultant staff should wear appropriate safety gear where required - helmets, reflectorized vests, footwear and goggles. First Aid kits and fire extinguishers are to be provided for all vehicles and offices. Staff members must wear seat belts in vehicles.

Consultant staff should be alert to possible dangerous situations on work sites, and provide warnings to Contractor's personnel.

Responsibilities

The DTL is responsible to provide a Visitor's Book on the site and record the name[s], date and time of all site visitors.

The DTL is responsible to ensure that sufficient safety equipment is available on site for all staff and visitors.

Site staff are to ensure that the Contractor's supervisors follow safe working procedures for workmen, that flagmen and/or traffic signs, markers or cones are deployed as necessary, and that water trucks are utilised to alleviate dust hazards, to ensure safe passage for civilian vehicles. Site staff should also ensure that the Contractor's supervisors pay particular attention to marking hazards at night(eg, excavations, broken down machinery, etc.).

2.23 Accident Report

Safety working practices to be observed at all time and the Contractor is required to have in place procedures that provide for safety both to his own work force and to the general public that could be contract with the work site.

Safety Plans are required to be submitted for the Engineer's approval to cover the Occupational Health and Safety of the workforce on site. Traffic Management Plans are also required for the Engineer's approval to provide for the safety of road users within the Contract Site.

These plans are to include measures that will be taken to avoid accidents. They must also include actions that are to be taken in lic that could be in contract with the work site.
 the event of an accident occurring.

In the event of an accident taking place, supervision staff are required to monitor the situation carefully to ensure that the procedures set out in the plans are being implemented as necessary.

Procedures and Responsibilities

The responsibility for safety rests with both the Consultant and the Contractor. The TL, the DTL and senior Engineers in charge of the particular facet of work are to ensure that the Contractor prepares and submits the requisite Safety Plans and Traffic Management Plans. These are to be approved by the TL.

If accidents do occur, then the following action is required from site staff:

2.24 Inspection Check Point

EARTH WORK(토공)

지형을 형성하고 구조물 기반을 조성하기 위한 절토, 성토, 정지 작업

주요 작업 항목	관련 문서 및 도면	품질관리 포인트
• Site Clearing and Grubbing • Excavation (cutting) • Embankment (filling & compaction) • Subgrade preparation • Slope protection and erosion control	• Earthwork Cross Sections • Mass Haul Diagram • Survey Baseline and Setting Out Plans	• Compaction Test (e.g. Field Density Test) • Moisture Content Monitoring • Joint Survey Records (Quantity Agreement)

DRAINAGE WORK(배수공)

노면 및 지반의 물 흐름을 제어하여 구조물 및 노면의 손상을 방지하기 위한 배수 체계 시공

주요 작업 항목	관련 문서 및 도면	품질관리 포인트
• Longitudinal & Cross Drain Construction • Culverts (Box, Pipe) • Open Channels and Side Ditches • Drain Inlets & Manholes • Subsurface Drainage Systems	• Drainage Plan & Profiles • Hydraulic Calculation Sheets • Manhole & Inlet Details	• Concrete Strength Test for precast units • Slope and Level Tolerances • Flow Test (if applicable)

BRIDGE AND STRUCTURE WORK(교량 및 구조물 공사)

교량, 옹벽 등 하중을 지지하거나 분산시키는 구조물의 시공

주요 작업 항목	관련 문서 및 도면	품질관리 포인트
• Foundation (Piling, Footings) • Substructure (Abutments, Piers) • Superstructure (Girders, Slabs) • Retaining Walls • Expansion Joints and Bearings	• Structural Drawings (Plan, Elevation, Reinforcement Details) • Bar Bending Schedule • Load Test Reports	• Reinforcement Check (spacing, type) • Concrete Pouring & Curing Control • Formwork Inspection

PAVEMENT WORK(포장공)

차량 주행을 위한 노면 구조 시공

주요 작업 항목
- Sub-base & Base Course (Granular, Crushed Stone)
- Asphalt Concrete Paving
- Joint Cutting and Sealing
- Surface Marking and Signage
- Sidewalks and Shoulders

관련 문서 및 도면
- Pavement Typical Sections
- Material Test Reports (Asphalt Mix Design, Aggregate Grading)
- Pavement Layer Thickness Records

품질관리 포인트
- Density and Compaction Test
- Asphalt Core Sampling
- Surface Regularity (using Straightedge or Laser)

TUNNEL WORK(터널공)

산악 또는 도심 구간 통과를 위한 암반 및 토사 지하 구조물 시공

주요 작업 항목
- Excavation & Support (NATM, TBM 등)
- Waterproofing Membranes
- Shotcrete and Rock Bolts
- Lining and Secondary Concreting
- Ventilation and Drainage Systems

관련 문서 및 도면
- Tunnel Cross Section Drawings
- Geotechnical Reports
- Support Pattern Diagrams

품질관리 포인트
- Convergence Monitoring
- Groundwater Control
- Shotcrete Thickness & Strength Test

ANCILLARY WORK(기타 공종)

기타 부대공사 또는 환경, 안전, 편의시설 설치 등

주요 작업 항목
- Road Furniture (Guardrails, Barriers)
- Landscaping and Topsoiling
- Sign Boards and Lighting
- Environmental Protection Measures
- Utility Relocation or Coordination

관련 문서 및 도면
- Landscaping Plan
- Utility Layout Drawings
- Road Safety Audit Reports

품질관리 포인트
- Material Compliance
- Alignment and Position Check
- Functionality Testing (Lighting, Signage)

2.25 Submittal and Approvals

1. C-ESMP

1. Introduction

2. Legal and Institutional Framework

2.1. Environmental Policies and Laws of Georgia

2.2. Operational Policy of World Bank

2.1. World Bank's Safeguard Policies

2.2. Georgian Legislation and the World Bank Requirements

3. Description of the Works

4. Sensitive Receptors and Environmental Values

4.1. Within project impact Zone

5. Environmental and Social Risk Assesment :

5.1. Environmental and social risk management

Table 5.1 Risk Assessment Matrix and Measures:

5.2. Historical and Cultural Heritage

6. Construction Environmental and Social Management Plan

6.1. Environmental Documents and Records

6.1. Environmental and Social Management Plan:

6.2. Environmental and Social Monitoring Matrix

6.3. Lay out Plan:

6.4. Environment Work Plan:

7. Social Management Plan

7.1. Social Specialist

7.2. Social Management Mechanism Scope of Responsibility

7.3. Social Management Report Mechanism

7.4. Resettlement Action Plan

7.4.1. Situation "0" Investigation Agency

7.4.2. Situation "0" of Property No.1

7.4.3. Situation "0" of Property No.2

7.4.4. Situation "0" of Property No.3

7.4.5. Situation "0" of Property No.4

7.4.6. Situation "0" of Property No.5

7.4.7. Situation "0" of Property No.6

7.4.8. Mechanism for Resettlement of Properties out of the Project

7.5. Safety Action Plan to Protect Local Communities

7.5.1. Description

7.5.2. General Action Steps

7.6. Hiring More Local Employees

7.6.1. CSCEC's Manpower Plan

7.6.2. Monitoring and Optimizing the Manpower Plan

7.6.3. Hire More Local Subcontractors to do Part of the Works

7.6.4. Hire local Foreman to Lead local Work Team;

8. Grievance Redress Mechanism Introduction

8.1. Objective

8.2. Grievance Redress Process

8.2.1. Receipt of grievances

8.2.2. Screening for Standing

8.2.3. Project-level Review of PAP Grievances

8.2.4. Grievance Redress Committee (GRC)

8.2.5. The Employer's level resolution (tier 3)

8.2.6. Closure of Grievances

8.2.7. Grievance log

8.2.8. The Contractor Contact Information

8.3. Workers Grievance Redress Mechanism,Code of Conduct

8.3.1. Labour-Related Grievance Mechanism

8.3.2. Code of Conduct (COC)

8.3.3. Grievance Submission Form

Annex 1. Noise Control Plan

1. Introduction
2. Scope and Objectives
3. Legislation
4. Baseline noise level
5. Sources of noise during pre-construction and construction
6. Mitigation and noise management measures
7. Monitoring
8. Responsibilities
9. Reporting
10. Revisions

Annex 2. Air Quality Management Plan

11. Introduction
12. Scope and Objectives
13. Legislation
14. Baseline Environment
15. Sources of impact
16. Description of impacts

17. Mitigation Measures

17.1. Dust Emissions

17.2. Air Pollutants and Greenhouse Gases

18. Monitoring

19. Monitoring locations

20. Responsibilities

21. Reporting

Annex 3. Construction Vibration Management Plan

22. Introduction

23. Objectives and tasks of vibration management plan

24. Condition Survey

25. Mitigation measures

26. Construction Vibration Management (CVMP)

27. Objective of vibration monitoring

28. Monitoring Location

29. Standards to be considered

30. Monitoring instruments

31. Monitoring process

32. References

Annex 4. Tunnel blasting vibration monitoring daily report.

2. Occupational Health and Safety(OHS) Management Plan

1. Introduction

   1.1. Purpose

   1.2. Occupational Health and Safety (OHS) Policy

   1.2.1. Statement of Commitment

   1.2.2. Implementation of Policy Commitment

   1.3. Definitions

   1.4. Responsibilities

   1.4.1. CSCEC-The Construction Company

   1.4.2. The Project Manager

   1.4.3. HSE Department-H.S.E. Manager

   1.4.4. HSE Department-HSS responsible person and Team

   1.4.5. Managers, Officers, Engineers and Supervisors

   1.4.6. Workers

   1.4.7. Contractors

   1.4.8. Visitors

   1.5. Consultation and Communication Arrangement

   1.5.1. Health and Safety Representatives (HSR)

   1.6. Training

   1.7. OHS Risk Assessment

   1.8. OHS Issue Resolution

   1.9. Authoritative Sources

   2. General OHS Information

   2.1. Emergency Procedures

   2.2. Hazard//Injury/Incident Reporting

   2.2.1. How to Report a Hazard or Injury or Incident:

2.3. Registering, Reporting and Inquiring of Notifiable Incidents

2.3.1. Hazard/incident/injury Registering-Reporting and Inquiring -

2.4. First Aid

2.4.1. First Aid - Summary for the H.S.E. Manager

2.5. OHS Training and Induction

2.5.1. Training

2.5.2. Documentation for Training

2.5.3. OHS Induction

2.5.4. Procedure

2.5.5. OHS Induction for Contractors/Visitors

2.5.6. Detailed OHS Induction for Contractors

2.6. Risk Management and the Risk Register

2.6.1. Risk Assessment

2.6.2. The Risk Management Process

2.6.3. Documentation for Risk Assessment

2.6.4. The OHS Risk Register

2.7. Workplace Hazard Inspections

2.8. Purchasing

2.9. OHS Record Keeping

2.10. Documents to Be Displayed

2.11. Important Contact Numbers

3. Specific OHS Requirements

3.1. Asbestos

3.2. Inappropriate Behaviour

3.3. Contractors

3.4. Dangerous Goods and Hazardous Substances

3.5. Electrical Safety

3.5.1. Residual Current Devices:

3.5.2. Unsafe Equipment:

3.6. Confined Spaces

3.7. Falls from Height

3.8. Manual Handling

3.8.1. Preventing Manual Handling injuries

3.8.2. When loading/unloading vehicles

3.9. Plant and Equipment

3.9.1. Risk Management

3.9.2. Maintenance and repair

3.9.3. Record Keeping

3.10. Personal Protective Equipment

3.11. Slips, Trips and Falls

3.11.1. Prevention

3.12. Drugs and Alcohol

3.13. UV Radiation

3.14. Vehicles

3.14.1. Alcohol and Drugs

3.14.2. Licenses

3.14.3. Mobile Phones

3.14.4. eat Belts

3.14.5. Smoking

3.14.6. Load Restraint in Vehicles

3.15. Working Alone

3.16. Blasting Works

3.16.1. Blasting Procedures

3.17. Excavations

3.18. Portable Tools

3.19. Inclement Weather

3.20. Fire Prevention

3.21. Permit to Work (PTW)

3.22. Confined Space Entry

3.23. Fuel and Oil Storage

3.24. Mobile Crane Safety

4. Forms and Checklists

4.1. Attachment 1 Emergency Contact List

4.2. Attachment 2 Hazard/Injury/Incident Registration

4.2.1. Submitted to relevant responsible person in the CSCEC

4.2.2. Submitted to Relevant State Supervision Authority in the Ministry of Internally Displaced Persons from the Occupied Territories, Labor, Health and Social Affairs of Georgia

4.2.3. Submitted to Relevant State Supervision Authority in the Ministry of Internally Displaced Persons from the Occupied Territories, Labor, Health and Social Affairs of Georgia

4.2.4. Submitted to relevant responsible persons in CSCEC

4.3. Attachment 3 OHS Induction Checklist for New Workers

4.4. Attachment 4 OHS Induction for Contractors/Visitors

4.5. Attachment 5 Detailed OHS Induction Checklist

4.6. Attachment 6 OHS Training Register

4.7. Attachment 7 OHS Risk Assessment Proforma

4.8. Attachment 8 OHS Hazard Inspection Procedure

4.9. Attachment 9 OHS Hazard Inspection Quick Checklist

4.10. Attachment 10 Suggested Asbestos Register

4.11. Attachment 11 Hazardous Substances Register

4.12. Attachment 12 Tech Requirements for Safe Working

3. Emergency Reponse Plan

1. Objectives and Tasks of the Emergency Response Plan

2. Brief Project Description

3. Types of Emergency Situations Expectedon

4. Characterization of the Potential Emergency Situation

4.1. Fire/Explosion

4.2. Salvo Spillage of Hazardous Substances including Oil Products

4.3. Personnel Traumatism and their Health & Safety Risks

4.4. Road Accidents

4.5. Natural Type Emergency Situations

5. General Preventive Measures for Emergency Situation

5.1. Estimated Scale of Incidents

5.2. Notification Scheme for Emergency Situation

6. Scheme of Emergency Response Organization

6.1. Response on Fire

6.2. Response on Explosion

6.3. Response on Accidental Spill of Hazardous Substances

6.4. Response to Traumatism of Personnel or Incident

6.4.1. First Aid during Fracture

6.4.2. First Aid during Wounds and Bleeding

6.4.3. First Aid in case of Burn

6.4.4. First Aid in Case of Electrical Trauma

6.5. Response on Traffic Accidents

6.6. Response on Emergencies of Natural Type

6.6.1. Response on Earthquake

6.7. Response on flood

6.8. Response during COVID 19 virus

6.8.1. Employer responsibility

6.8.2. Obligation of employees

6.8.3. Employees who should not arrive at workplace

6.8.4. Requirements for project transport department

6.8.5. Action plan in case of detection related symptoms

7. Personnel and Equipment Required for Emergency Response

7.1. Personnel Required for Emergency Response

7.2. Equipment Required for Emergency Response

8. Necessary Qualification and Personnel Training

9. Monitoring and Reporting

Appendix

A.
Document Tracking Forms

FORM: DT-1

DOCUMENT TRACKING LOG SHEET

Doc.No.	Ref. No.	From	To	Subject/Description	Received By			Released By			Remarks
					Name	Initial	Date	Name	Initial	Date	

FORM: DT-2

INCOMING CORRESPONDENCE LOG

Letter No.	Date Written	Date Received	Subject	File

FORM: DT-3

OUT CORRESPONDENCE LOG

Letter No.	Date Written	Date Sent	Subject	Remarks

FORM: DT-4

VISITORS REGISTER

Date	Visitor's	Company	Person Visited	Purpose of Visit	Time-in	Time-Out

B.
Construction Supervision Forms

B.1
Correspondence/ Communication Forms

FORM: CC-1

NOTICE OF MEETING

Date :
To   :

What	:	_____	_____
When	:	_____	_____
Where	:	_____	_____
Time	:	_____	_____

The attendance of requested to discuss the following,
Where applicable, sketches, measurements, calculations, DWG Test Sheet etc to be provided in this space

:   :

CC : _____
       _____

FORM: CC-2

MINUTES OF COORDINATING MEETING NO. _____

Date/Time   : _____
Location   : _____
Attendees   : _____

A. Employer

  _____
  _____

B. Engineer

  _____
  _____

C. Contractor

  _____
  _____

Subject Matters Discussed:

Item No.	Minutes	Action	
		By Whom	By When

:

_____
Employer

_____
Consulting Engineer

_____
Contractor

FORM: CC-3

Meeting _____

ATTENDANCE SHEET

Date/Time _____
Venue _____

	Names	Position	Company/Agency	Signature
1				
2				
3				
4				
5				
6				
7				
8				
9				
10				
11				
12				
13				
14				
15				
16				
17				
18				
19				
20				
21				
22				
23				
24				
25				
26				
27				
28				
29				
30				
31				
32				
33				
34				
35				

FORM: CC-4

INTER-OFFICE CORRESPONDENCE

Date   :

To   :

From   :

Subject :

FORM: CC-5

SPEED LETTER

Date _____
To _____   From _____
   _____   _____

Attention: _____

_____
_____
_____
_____
_____
_____
_____
_____
_____
_____
_____

                              Signed: _____

Reply _____
_____
_____
_____
_____
_____
_____
_____
_____
_____
_____
_____

                              Signed: _____

Appendix

FORM: CC-6

COMMUNITY RELATIONS CONTACT SHEET

Contact Name: _____
Address
_____
Day Telephone _____ Night _____
Affiliation / Community Organization _____
Phone for Contact: _____
_____
_____
_____
_____
_____
_____

Referred by _____
Title _____ Date _____
Location _____
Construction Site / Community _____
Photo Taken: _____
_____
_____
_____
_____
_____
_____

Signature _____ Date _____

B.2
Supervision Forms

FORM: CS-1

WORK REQUEST FORM

To:(Consultant)	Requested Work Scheduled To Start On: Date: _____ Time: _____
From:(Contractor)	Note: Contractor to submit request in triplicate a Minimum of 48 hours in advance of scheduled start
Item No.	Item Description
Description of Work	
Location	
Equipment to Be Used	
Estimated Quantity to Be Accomplished	
Submitted by: _____ Contractor	Received by: _____ Date: _____ Time: _____ Office of the Resident Engineer
Inspected by:	**Comments(Initiated and Dated):**
Quality Assurance Engineer	
Surveyor	
Inspector	
☐ Request ☐ Approved ☐ Disapproved _____ Team Leader	Accepted by: _____ Contractor Date: _____ Time: _____

FORM: CS-2

MATERIAL INSPECTION FORM

To _____
From _____
Date _____
Material for which inspection is requested:
_____

Item No.	Item Description	Location	Requested Date	Time	Remarks

Results of Inspection _____
_____

Requested by: Received by:

_____   _____
       Contractor   Quality Assurance Engineer

Inspected by: Comments:

_____   _____

Date:

Note: Material is

a. Contractor to submit request a   Approved/Disapproved by:
   minimum of 48 hours in advance
b. Consultant to return approved or   _____
   disapproved original and one copy   Resident Engineer
   to the Contractor before the work
   proceed.

Appendix

FORM: CS-3 (Page 1 of 2)

INSPECTION CHECKLIST "EXAMPLE"
(To Be submitted with WRF)

Division No.6: Concrete Works   Date of Inspection: _____

Structure:

(Location) _____   Station/Location: _____

A. Test	Complying	Non-Complying	Initial
1. Beam/Cylinder Molds			
2. Curing Materials			
3. Slump Cone			
4. Plumbness			
5. Quality of Materials Used			
B. Line and Grade			Initial
1. Form Setting			
2. Elevation			
3. Vertical Alignment			
4. Horizontal Alignment			
5. Width			
6. Thickness			
7. Others			
C. Structural			Initial
1. Bar Size			
2. Bar Alignment			
3. Depth of Member			
4. Length of Member			
5. Bearing Pads			
6. Lapping Bars			
7. Anchor Bars			
8. Formwork			
9. Vertical Alignment			
10. Bar sizes, spacing and number			
11. Bar installation requirements			
12. Others			

FORM: CS-3 (Page 2 of 2)

D. Materials&Equipment	Complying	Non-Complying	Initial
1. Cement			
2. Fine Aggregates			
3. Course Aggregates			
4. Water			
5. Mixer			
6. Concrete Vibrator			
7. Sampling Molds, Cylinders, etc.			
8. Protective Covering Materials			
9. Concrete Saw			
10. Concrete Block Spacers			
11. Finishing Tools Equipment (Screeder/Broom, etc.)			

Inspected/Checked by:

_____   _____   _____
Road Engineer   Bridge Engineer   Quality Assurance
Consultant   Consultant   Engineer Consultant

Arrpoved/Disapproved by:

_____
Team Leader

FORM: CS-4

SITE INSTRUCTION SI NO._____

To   : _____
SUBJ   : _____
Date   : _____

Observation/Comments:
_____
_____
_____

Defects:
_____
_____
_____

Instructions:
_____
_____
_____

Issued by: Received by: Approved by:

_____   _____   _____
     Consultant   Contractor   Team Leader

Note:
A formal letter confirming this Site Instruction from the Engineer will follow within 48 hours.

FORM: CS-5

REQUEST FOR INFORMATION(RFI)

Date Requested: _____	Control: RFI-_____
	Page___of_____
Request Addressed to: _____	RFI No.: _____
RFI From: _____	Contract No.: _____
Subject: _____	

Description of Information Needed(include reference):

Received: _____ Date: _____
               Resident Engineer

Reply:

Replied by: _____ Approved by: _____
                                                             Team Leader

Received: _____ Date: _____
               Contractor

Distribution:
   ___Team Leader   ___Contractor
   ___Resident Engineer   ___Document/Record
   ___Officer   ___Other

FORM: CS-6

SUSPENSION ORDER NO. _____

To: _____   Location: _____
                                     Date   : _____

You are hereby directed to suspend all construction/operations on the above project, _____ effective on the *Day* of *the Month* day of *Month of the Year,* *Year* because of _____ Until such time a Resume Order is issued.

Contract Time including Extension   _____days
Days Used to Date   _____days
Day Remaining   _____days

Please acknowledge receipt of this, dating, signing, and returning two of the attached copies. The third copy is for your file.

By:

_____
Resident Engineer

Approved:

_____   _____
Team Leader   Date

I hereby acknowledge receipt of the above notice.

Date: _____   By: _____
                                     Contractor

FORM: CS-7

RESUME ORDER NO.____

To: _____ Location: _____

　　　　　　　　　　　　　　　Date   : _____

    You are herby directed to suspend all construction/operations on the above project, _____ effective on the *Day   of   the Month* day of *Month of the Year,   Year.*

    Please acknowledge receipt of this, dating, signing, and returning two of the attached copies. The third copy is for your file.

Please Be guided accordingly.

　　　　　　　　　　　　　　　　　　　　By:

　　　　　　　　　　　　　　　　　　　　_____

　　　　　　　　　　　　　　　　　　　　Resident Engineer

　　　　　　　　　　Approved:

_____　　　　　　　_____

Team Leader　　　　　　　　　　　　　　Date

I hereby acknowledge receipt of the above notice.

Date: _____　　　By: _____

　　　　　　　　　　　　　　　　Contractor

FORM: CS-8

INSPECTOR'S DAILY REPORT

Segment/Structure:　　　　　Report No.:　　　　　Date:　　　　Shift:
Segment/Structure Supervisor:　　　　　　　　　Water Level　　A.M.:
　　　　　　　　　　　　　　　　　　　　　　　　　　　　　　P.M.:

Manpower(see code at the back)								Activities	
10	20	30	40	50	60	70	80	Total	Pay Item	Description of Work including Quantity
								Total		

Plant/Equipment on Site				Number of Hours		
Description	QTY	Operation	Deliveries	Oper	Idle	Down

FORM: CS-9

DAILY ACCOMPLISHMENT SUMMARY REPORT

Date/Day
Weather   A.M._____
           P.M._____

Activities

Item No.	Description/ Location	Unit	Accomplishment			Remarks
			PREV.	Today	To Date	

Prepared: Checked by: Noted by:

_____   _____   _____
Project Inspector   Resident Engineer   Team Leader

FORM: CS-10

DAYWORK REQUEST

Date: _____

No: _____

Contractor : _____

Station Limit : _____

Description of Work : _____

Equipment to Be Used	Manpower to Be Utilized	Materials to Be used

Scheduled Start of Work : _____

Time to Date

Note:

This Request must be submitted and approved at least one day prior to actual start of work

Checked by:

_____   _____   _____
Road Engineer   Bridge Engineer   Quality Assurance Engineer

Recommending Approval: Approved by:

_____   _____
Resident Engineer   Team Leader

FORM: CS-11

ACCOMPLISHMENT FOR DAYWORK

Name of Project:

Date _____
Project _____

Contractor _____

Work
Description

Equipment	Number	No. of HRS	Hourly Rate	Amount
			Total	

Labor	Number	No. of HRS	Hourly Rate	Amount
			Total	

Equipment	Unit	Quantity	Hourly Rate	Amount
			Total	

Description	
A. Today's Equipment Cost	P
B. Today's Labor Cost	P
C. Today's Material Cost	P
D. Total Cost of Today's Work (A+B+C)	P

We certify that the above labor was performed, the equipment and material were used.

Submitted by:

_____
Team Leader Contractor

Checked by:

_____   _____
Resident Engineer   Representative
Consultant   EMPLOYER
Noted by: Approved by:

_____   _____
Team Leader   Representative
Consultant   EMPLOYER

Appendix

FORM: CS-12

WEEKLY WORK SCHEDULE
(Inclusive Date) From _____ to _____
Sheet _____ of _____

| Item of Work | Prev. Accomp. | Quantity | | UNIT | Sun | Mon | Tue | Wed | Thu | Fri | Sat | Remark |
		This week	To date		1	2	3	4	5	6	7	
1												
2												
3												
4												
5												
6												
7												
8												
9												
10												

Submitted by: Verified by: Approved by:

_____   _____   _____
Contractor   Site Engineer   Resident Engineer

FORM: CS-13

MONTHLY WORK SCHEDULE
For the Month of ____
Sheet ___ of ___

Item of Work	Quantity	Unit	1	2	3	4	5	6	7	8	9	10	11	12	13	14	15	2010 16	17	18	19	20	21	22	23	24	25	26	27	28	29	30	31	Remark
1																																		
2																																		
3																																		
4																																		
5																																		
6																																		
7																																		
8																																		
9																																		
10																																		

Submitted by: Verified by: Approved by: Noted by:

_____   _____   _____   _____
Contractor   Site Engineer   Team Leader   EMPLOYER

FORM: CS-14

MANPOWER STATUS REPORT

For the Month of _____

Number-Scheduled/Actual

Manpower	1	2	3	4	5	6	7	8	9	10	11	12	13	14	15	16	17	18	19	20	21	22	23	24	25	26	27	28	29	30	31	Total (Man-days)	Remark

Submitted by:

_____
Contractor

Verified by:

_____
Resident Engineer

FORM: CS-15

EQUIPMENT STATUS REPORT
For the Month of ____

| Equipment | \multicolumn{32}{c}{Number-Scheduled/Actual} | Remark |

Equipment	1	2	3	4	5	6	7	8	9	10	11	12	13	14	15	16	17	18	19	20	21	22	23	24	25	26	27	28	29	30	31	Total	Remark

Submitted by: _____ Verified by: _____

Contractor   Resident Engineer

Appendix 151

FORM: CS-15a

CONTRACTOR'S MANPOWER

Contractor:_____   Date:_____

Position /Designation	No. Required	No. On Site	No. of Days				Remarks
			Rest	Present	Leave	Absent	

Submitted by: Verified by:

_____   _____
   Contractor   Resident Engineer

FORM: CS-16

JOINT QUANTITY MEASUREMENTS

Work: _____
Section/Area: _____ Bill no.: _____ Page _____ of _____
Date: _____ Day of Week: _____ Time: _____

Pay Item No.	Description of Item	Unit	Quantity Accomplished					Measured by		
			Not Yet Tested	Failed Test	Passed Test	Total	Engineer Representative	Contractor Representative	BCDA SCD/PMO	Remarks

Checked by: _____   Approved by: _____

    Contractor   Resident Engineer

FORM: CS-16a

CONTRACTOR'S MANPOWER

Contractor:_____   For the period:_____

Designation	Min No. Required	No. On Site	Equip. No.	Status(C.D.)			Remarks
				OPL	BD	IDLE	

Submitted by: Verified by:

_____   _____
    Contractor   Resident Engineer

FORM: CS-17

MONITORING SHEET

Type of Structure/Construction Activity	Station/Location of Measurement	Dimension (As per Plans/Specs)	Dimension (As Measured Actual)	Variation (+/-)	Tolerance	Judgment (Pass/Fail)

FORM: CS-18

PROJECT LOGBOOK FORMAT

Location : _____
Date : _____

1. Weather, specifically if relevant to work progress
    - Period and intensity of rain
    - Approximate direction and strength of wind
2. Work Force
    - Absence of Key personnel
    - Starting and finishing times of workday particularly during critical periods of the work
3. Activities
    - New Starts
    - Completions
4. Inspection performed during the day
    - Location and description of work inspected
    - Remarks concerning unusual features of the work
    - Tests/Results
    - Defective work to be corrected later
5. Significant delays and their causes
6. Accidents
7. Instructions received and details of instructions or warnings given to or important conversations held with eh Contractor or its representative.
8. Delivery or removal of equipment, plant or materials critical to the work program.
9. Problems encountered and actions taken.
10. Visits of key personnel from BCDA and Contractor noting important matters arising from such visits.
11. Relevant incoming or outgoing radio/telephone calls; noting time, caller and subject
12. Compliant received from residents of area or the general public.
13. Photographs to illustrate poor workmanship satisfy hazards, work progress and site conditions, time and place they were taken recorded and cross-reference with project logbook.
14. Remarks.

Prepared by   Concurred by

_____   _____
          Site Engineer   Resident Engineer

FORM: CS-19

REQUEST FOR OVERTIME

This is to inform you that we will render over time on _____
(Segment)

from _____ to _____ on _____
(Time Start)   (Time End)   (Date)

In this connection we would like to request your representative to be present at the aforesaid jobsite on the date and time started.

_____
Contractor

Remarks:

_____
_____
_____

Approved/Disapproved by:

_____
Resident Engineer

Noted by:

_____
Team Leader

C.
Contract Administration Forms

FORM: CA-1

Date: _____

INTERIM PAYMENT CERTIFICATE NO. ___

For the Period of:
Loan Agreement ID:
Name of the Project:
Name of the Contractor:
Original Contract Amount:

Currency	Local Currency (USD)	Foreign (USD)
1. Total Value of Work to Date	_____	_____
2. Total Value of Work/Billing in the Previous Period	_____	_____
3. Total Value of Work/Billing for this Period	_____	_____
4. Deductions		
A. 5% Retention	_____	_____
B. 30% Recoupment from Advance Payment	_____	_____
C. 10% Value Added Tax(VAT)	_____	_____
Total Deduction (A+B+C)	_____	_____
5. Amount Approved for Payment	_____	_____
6. Add: Cost of Materials Previously Delivered	_____	_____
7. Less: Repayment of Materials Previously Delivered	_____	_____
8. Net Amount due to Contractor	_____	_____
Scheduled Physical Progress to Date:	: _____	: _____
Actual Physical Progress to Date:	: _____	: _____

Advance(+) Slippage(-)   : _____
Contract Completion Date : _____

9. I certify that this Interim Payment Certificate is correct, and that no payment for same has been received. With respect to any payments included for materials, said materials are on hand, are properly stored and protected at or near the site of the work, have been inspected and approved, are not in excess of the estimated quantities required, and are to be incorporated into this project. The material invoice price plus freight are the net price paid, less all discounts and rebates, and there are no liens on said materials.

Contractor's Representative

10. I certify that I have checked the quantities covered in this certificate, that the work was actually performed, that the quantities, including materials furnished but not incorporated in the work, are correct and consistent with all previous recommendations for payment and that the work, quantities and amount are consistent with the requirements of the Contract.

11. Certificate Recommended for Payment

| Team Leader | Representative |
| Consultant | EMPLOYER |

12. Certificate Approved for Payment

| Deputy Director, EMPLOYER |
| Deputy Chairman, EMPLOYER |

FORM: CA-2

STATEMENT OF WORK ACCOMPLISHED

INTERIM PAYMENT CERTIFICATE NO. _____
FOR THE PERIOD OF _____, 20 _____

Statement Summary

Description	Formula	Total Amount		Until Previous Period		This Period	
		Local(USD)	Foreign(USD)	Local(USD)	Foreign(USD)	Local(USD)	Foreign(USD)
Construction Cost							
Total Construction Cost	A						
Price Escalation	B						
Daywork	C						
Variation Order							
- Change Order	D						
- Variation Order	E						
- Extra Work Order	F						
Total Gross Amount	G=A+B+C+D+E+F						
Deductions							
Retention Money	H=G*5%						
Repayment of Advance Payment VAT	I=G*30%						
Payment of Materials on Site This Period	J						
Total Deduction	K=H+I+J						
Total Amount After Deductions	L = G-K						
Materials Delivered on Site This Period	M						
Net Amount of This Statement	N=L+M						

Progress Monitoring	Total Gross Amount(%)	
	After Deduction Amount(%)	

Total Amount of Contract in Local Currency

Net Amount in Words
Local Currency = USD: _____
Foreign Currency = USD: _____

Note: Sub-Clauses mentioned above refer to particular provisions in FIDIC - (First Edition 1999). Conditions of Particular Application - Volume II and Addendum.

Submitted by: Revised by: Certified by:

Contractor   Team Leader   Representative EMPLOYER
Date: Date: Date:

FORM: CA-3

STATEMENT OF WORK ACCOMPLISHED
INTERIM PAYMENT CERTIFICATE NO. _____
FOR THE PERIOD OF _____ , 20 _____

Summary by Work Division

| Division No. | Description | Contract Amount ||||| To Date Amount || Previous Amount || This Amount ||
| | | Local Amount(USD) || Foreign Amount(USD) || Local (USD) | Foreign (USD) | Local (USD) | Foreign (USD) | Local (USD) | Foreign (USD) |
		Sub-total	Total	Sub-total	Total						
1	General Requirement										
2	Site Work										
3	Earth Work										
4	Waterway Work										
5	Asphalt Product and Pavement										
6	Concrete Work										
7	Steel Work										
8	Road Furniture and Miscellaneous										
9	Day Work										
	Total Cost										

Submitted by: _____ Revised by: _____ Certified by: _____

     Contractor   Team Leader   Representative EMPLOYER

Date: _____   Date: _____   Date: _____

FORM: CA-4

STATEMENT OF WORK ACCOMPLISHED
INTERIM PAYMENT CERTIFICATE NO. _____
FOR THE PERIOD OF _____, 20___

Summary of Progress

Pay Item No	Description	Unit	Contract Quantity (A)	Unit Rate		To Date			Until Previous Period			This Period			Percentage(%)			Remarks
				Local (USD) (B)	Foreign (USD) (C)	Quantity (D)	Local (USD) (E)=(B)×(D)	Foreign (USD) (F)=(B)×(C)	Quantity (G)	Local (USD) (H)=(B)×(G)	Foreign (USD) (I)=(C)×(G)	Quantity (J)=(D)-(G)	Local (USD) (K)=(E)-(H)	Foreign (USD) (L)=(F)-(I)	To Date (M)=(D)/(A)	Previous (N)=(G)/(A)	This Period (O)=(M)/(N)	
Total																		

Submitted by: _____   Revised by: _____   Certified by: _____   Approved by: _____

Date: _____   Date: _____   Date: _____   Date: _____

Contractor   Engineer   Team Leader   EMPLOYER

FORM: CA-5

QUANTITY CALCULATIONS SUMMARY BY PART
BILL2
PART E
INTERIM PAYMENT CERTIFICATE NO.___
FOR THE PERIOD OF _____

Pay Item	Description	Original Contract Quantity		Progress Quantity			Remarks
		Unit	Quantity	To Date	Previous Period	This Period	

Prepared by:

Conducted an Inspection and Verification

_____

Contractor's Quantity Surveyor

_____

Engineer's Material Inspector

_____

Engineer's Quality Assurance Engineer

_____

Team Leader

FORM: CA-6

QUANTITY CALCULATIONS SEGMENT/PART
BILL2
PART E
INTERIM PAYMENT CERTIFICATE NO.___
FOR THE PERIOD OF _____

Pay Item No. and Description	Location	Unit	Progress Quantity			Remarks
			To Date	Previous Period	This Period	

Prepared by: Conducted an Inspection and
                                              Verification

_____   _____
Contractor's Quantity Surveyor   Engineer's Material Inspector

                                              _____
                                              Engineer's Quality Assurance Engineer

                                              _____
                                              Team Leader

FORM: CA-7a

DETAILED QUANTITY CALCULATION(ACCOMPLISHMENT)

Consultant:	Month:
	Monthly Certificate No:
	Contractor:
Bill No.:	Original Quantity:
Pay Item No.:	Revised Quantity:
Description:	
Unit:	
This Certificate, Quantity:	
Cumulative Quantity:	
We certify that unacceptable work has been deducted, and not included in this calculation.	
Submitted By: _____ Contractor	Conducted an Inspection and Verification: _____ _____ Site Engineer Site Engineer Approved By: _____ Team Leader Noted By: _____ EMPLOYER

FORM: CA-7b

DETAILED QUANTITY CALCULATION(ACCOMPLISHMENT)

Contractor:	Month:
	Monthly Certificate No.:
	Contract No.:

Bid No.:	Original Quantity
Pay Item No.:	Revised Quantity
Description: Reinforcing Steel	
Unit: kg.	

Bar Bending Diagram

Type of Structure	Bar Mark	Bar Size (mm)	Bar Shape	Quantity (pcs)	Dimensions(mm)						Length Per Bar (mm)	Total Length (mm)	Unit Wt (kg)	Total Weight(kg)
					a	b	c	d	e	f				
													Total	

This Certificate, Quantity:
Cumulative Quantity:
We certify that unacceptable work has been deducted, and not included in this calculation.

Submitted by:	Conducted an Inspection and Verification:
_____	_____ _____
Contractor	Site Engineer Contractor Site Engineer
	Approved by:
	_____
	Team Leader
	Noted by:
	_____
	EMPLOYER

FORM: CA-7c

DETAILED QUANTITY CALCULATION(ACCOMPLISHMENT)

Consultant:	Month:
	Monthly Certificate No:
Package No./Section:	Contractor:

Bill No.: Original Quantity:
Pay Item No.: Revised Quantity:
Description:
Unit:

Station	Distance	Travelled Way	Area(sq.m.)		Total Area	Volume (cu.m.)
			Shoulder			
			Left	Right		
					Total	

This Certificate, Quantity:
Cumulative Quantity:
We certify that unacceptable work has been deducted, and not included in this calculation.

Submitted By:	Conducted an Inspection and Verification:
_____ Contractor	_____ _____ Site Engineer Site Engineer Consultant Consultant Approved By: _____ Team Leader Noted By: _____ EMPLOYER

Appendix 167

FORM: CA-7d

DETAILED QUANTITY CALCULATION(ACCOMPLISHMENT)

Consultant:	Month:
	Monthly Certificate No:
	Contractor:

Bill No.:	Original Quantity:
Pay Item No.:	Revised Quantity:
Description:	
Unit:	

| Station | Distance | Travelled Way | Area(sq.m.) | | | Density | Volume (cu.m.) |
| | | | Shoulder | | Total Area | | |
			Left	Right			
					Total		

This Certificate, Quantity:

Cumulative Quantity:

We certify that unacceptable work has been deducted, and not included in this calculation.

Submitted By:	Conducted an Inspection and Verification:	
Contractor	Site Engineer Consultant	Site Engineer Consultant
	Approved By:	
	Team Leader	
	Noted By:	
	EMPLOYER	

FORM: CA-7e

DETAILED QUANTITY CALCULATION(ACCOMPLISHMENT)

Consultant:	Month:
	Monthly Certificate No:
	Contractor:

Bill No.:	Original Quantity:
Pay Item No.:	Revised Quantity:
Description:	
Unit:	

Station	Distance	Soil Type	Area(sq.m.)			Depth	Area (sq.m.)	Volume (cu.m.)
			Left	Right	Total			
							Total	

This Certificate, Quantity:

Cumulative Quantity:

We certify that unacceptable work has been deducted, and not included in this calculation.

Submitted By:	Conducted an Inspection and Verification:
_____	_____ _____
Contractor	Site Engineer Site Engineer
	Consultant Consultant
	Approved By:
	_____
	Team Leader
	Noted By:
	_____
	EMPLOYER

Appendix

FORM: CA-8

INTERIM PAYMENT CERTIFICATE No.: _____
For the Period of ____ to ____, 20____

STATEMENT OF TIME ELAPSED AND WORK ACCOMPLISHED

Project ID   : EWCIP Chumateleti-Khevi Section

Contractor   : _____
Period Covered   : _____
Original Contract Time   : _____
Date of Effectivity of Contract   : _____
Revised Contract Amount Due
To Approved Change Order/
Extra Work Order   : _____
Original Expiry Date   : _____
Approved Time Extension   : _____
Shutdown Days Due to
Suspension of Work   : _____
Total Time Extension   : _____
Total Calendar Day's Elapsed
To Date   : _____
Revised Expiry Date Due to
Time Extension   : _____
Percentage of Time Elapsed   : _____
Schedule of Work Accomplished: _____
Advance(Slippage)   : _____

Certified by:

_____
Team Leader

FORM: CA-9

EMERGENCY WORK ORDER

Emergency Work Order No.: _____

To: _____ Date: _____

You are here by directed to perform the following extra work not included in the plans for this contract on an immediate basis. A formal change order will be prepared, in due course.

Detailed Description of Emergency Work Order:
_____
_____
_____
_____

Method of Payment shall be as follows:
_____

Accepted :
_____
          (Contractor) (Date)

Recommended :
_____
          (Engineer) (Date)

Issued by :
_____
          (Engineer) (Date)

cc: EMPLOYER
    Team Leader

FORM: CA-10

VARIATION ORDER

Contractor: _____ Variation Order No. _____
NOTE: This Variation Order is not effective until approved by _____
(Authorized EMPLOYER Official)

Of the Road Department of MINISTRY of REGIONAL DEVELOPMENT AND INFRASTRUCTURE

I. Change Requested by
_____

II. In accordance with General Conditions of Contract, Clause _____ you are hereby directed to perform the following work:
_____
_____
_____
_____

III. Payment for this work will be at the rates attached below:
IV. By Reason of this Variation Order:
      1. The time for completion will be increased / decreased by
          _____ calendar days.
      2. The contract amount (Price) is increased / decreased by
          _____ Pesos.

V. We, the undersigned Contractor have given careful consideration to the change proposed and hereby agree, if this Variation Order is approved, that we will provide all equipment, furnish all materials, and perform all service necessary for the satisfactory completion of the work specified herein. We will accept as full payment thereof the process shown herein.

Signed with/without reservation:

Accepted   : _____   Date _____
                                 Contractor

VI. Endorsed by   : _____   Date _____
                                 Team Leader

VII. Endorsed by   : _____   Date _____
                                 EMPLOYER

FORM: CA-11

Month of _____

Contractor: _____ Monthly Payment Certificate No. _____

STATEMENT OF MATERIALS ON SITE

Amount of materials on site from previous monthly payment

_____
_____

We hereby certify that the following materials were delivered this period at the Project Site.

Description	Unit	Unit Price	Quantity	Amount	Remarks
Total					
80% of Total					

Note:
1. The amount to be credited to the contractor shall be the equivalent of 70% of the Contractor's reasonable cost of the materials and Plant delivered on Site as determined by the Engineer in accordance with the Particular Conditions of Contract.
2. Bank Guarantee issued by _____, on _____, 2010 is hereto attached.

Certified by:

_____ _____ _____
Quality Assurance   Site Manager   Quality Assurance
Contractor   Contractor   Engineer Consultant

Noted by:

_____
Team Leader

FORM: CA-12

MONTHLY STATEMENT NO. _____
FOR THE MONTH OF _____, 20 _____

MATERIALS ON SITE SUMMARY REPORT

BOQ Item No.	Description	Unit	Unit Rate		Quantity Delivered (Inspected and Approved)			Quantity Consumed (Based on Inspection Sheet)			To Date Total Remaining Quantity on Site	Amount to Be Credited (Amount of Materials Delivered This Month)		Amount to Be Debited (Amount of Materials Consumed This Month)		Remarks
			Local (USD)	Foreign (USD)	Total To Date	Until Previous Month	Delivered This Month	Total To Date	Until Previous Month	Delivered This Month		Local (USD)	Foreign (USD)	Local (USD)	Foreign (USD)	
Total Amount																
Amount to Be Carried to Statement Summary (70% of the Total Amount)																

In Words:
Amount to Be Credited:
Local Currency _____ USD: _____ Foreign Currency _____ USD: _____
Amount to Be Debited:
Local Currency _____ USD: _____ Foreign Currency _____ USD: _____

Submitted by: Received by: Recommending Approval:

     Contractor   Site Engineer Consultant   Team Leader
     Date: _____   Date: _____   Date: _____

FORM: CA-13

MONTHLY STATEMENT NO._____
FOR THE MONTH OF _____, 20____

MATERIALS ON SITE DELIVERY AND CONSUMPTION REPORT

Materials Delivered This Month

Item No.	Description	Unit	Quantity	Delivery Date	Remarks

Materials Consumed This Month

Item No.	Description	Unit	Quantity	Remarks
				Please Refer to Part _____ Summary
				Please Refer to Part _____ Summary
				Please Refer to Part _____ Summary
				Please Refer to Part _____ Summary
				Please Refer to Part _____ Summary
				Please Refer to Part _____ Summary
				Please Refer to Part _____ Summary
				Please Refer to Part _____ Summary
				Please Refer to Part _____ Summary
				Please Refer to Part _____ Summary

Checked by:

_____   _____
          Site Engineer   Site Engineer

Certified by:

_____
          Team Leader

FORM: CA-14

REQUEST FOR PAYMENT OF MATERIALS-ON-SITE
For the Month of _____

Contractor: _____

In accordance with the provision of the Contract Documents, the Contractor is requesting as stated on his letter dated _____ for the payment of "Material - On - Hand" for the following materials:

Materials Description	Quantity	Units	Invoice Cost or Materials Cost Per Bid Price*	Where Stored
Total				
80% of Total				

*Whichever is lower

AFFIDAVIT

The materials listed above have been purchased exclusively for use on the above referenced project. The material is separated from other like materials and is physically identified as our property for use only for CHUMATELETI-KHEVI PROJECT _____. The Engineer may enter upon premises where they are stored for the purpose set forth in the contract documents for inspection, checking, auditing or for any other purpose as maybe considered necessary. It is expressly understood and agreed that this information and affidavit is furnished by the for the purpose of obtaining payment for the above materials before they are incorporated into the Works. Storage thereof of the locations shown is under the sole responsibility of the Contractor.

Submitted by: Verified by:

_____   _____
Contractor   Team Leader

Noted by:

_____
EMPLOYER

FORM: CA-15 (Page 1 of 2)

REQUEST FOR PROPOSAL(RFP) NO. _____

Part. I Employer's Use

To: _____ Site Manager, Contractor	From: _____ Representative, EMPLOYER	Date:

Please be informed of the changes indicated below which are being considered for inclusion under this Contract. We would appreciate if you can estimate the cost involved and submit the accomplished estimate to us not later than _____ for evaluation. This notice is only a request for a proposal and does not constitute authorization to proceed with the Work.
Description of Changes:
Scope 1
Scope 2
Scope 3
Reason for the Changes in Scopes ___, ___, ___:
Specification No.'s or Drawing No.'s Affected
Remarks:

FORM: CA-15 (Page 2 of 2)

Part. II Contractor's Use

To: EMPLOYER	From:	Date:
Thru: Team Leader, Engineer	_____ Site Manager, Contractor	

We have accomplished the estimate Form for the changes below. The effect on the Schedule and the cost are summarized below. For your ready reference, the estimate Form/Construction Schedule (CPM) are attached.

We will appreciate it if you can evaluate this change soonest.

Contractor's Proposed Time Change Scope 1: _____ day(s) USD _____ Scope 2: _____ day(s) USD _____ Scope 3: _____ day(s) USD _____	Evaluated Time Change Scope 1: _____ day(s) Scope 2: _____ day(s) Scope 3: _____ day(s)
Remarks:	
Recommended Amount Scope 1: _____ day(s) USD _____ Scope 2: _____ day(s) USD _____ Scope 3: _____ day(s) USD _____	Recommended Time Change _____ day(s) _____ day(s) _____ day(s)
Original Contract Amount Deductive Contract Amount Due to RFP No. _____ Contract Amount After This Change Period of Completion Prior to This Change Additional Contract Time Due to RFP No. _____ Period of Completion After This Change	USD USD USD _____ Calendar days _____ Calendar days _____ Calendar days
Reflected reconciled amount is agreed upon, by and between the Contractor, _____ and the **EMPLOYER**	

Part. III Employer's Use

To: _____ Site Manager, Contractor	From:	Date:
Thru: _____ Team Leader, Engineer	_____ Site Manager, Contractor	

Connection with our Contract for the construction of EWCIP Chumateleti-Khevi Section, Package ____:_____ Section, you are directed to provide all labor materials, equipment and supervision necessary to complete the changes described above. This will be the basis of Payment for your accomplishment to the Project. This new becomes RFP No. ____.

☐ Approved ☐ Cancel Changed ☐ Renegotiate

FORM: CA-15a (Page 1 of 2)

REQUEST FOR PROPOSAL(RFP) NO._____

Part I Employer's Use

To: _____ Site Manager, Contractor	From: _____ Representative, EMPLOYER	Date:
Please be informed of the changes indicated below which are being considered for inclusion under this Contract. We would appreciate if you can estimate the cost involved and submit the accomplished estimate to us not later than _____ for evaluation. This notice is only a request for a proposal and does not constitute authorization to proceed with the Work.		
Description of Changes:		
Scope 1		
Scope 2		
Scope 3		
Reason for the Changes in Scopes _____, _____, _____ :		
Specification No.'s or Drawing No.'s Affected		
Remarks:		

Part II Contractor's Use

To: _____ Deputy Director, EMPLOYER To: _____ Team Leader, Consultant	From: _____ Site Manager, Contractor	Date:

We have accomplished the Estimate Form for the above change. The effect on the Schedule and the Costs are summarized below. For your ready reference, the Estimate Form/CPM Diagram are attached.
We appreciate if you can evaluate this change request.

Time Change: Amount:
☐ Add. Calendar Days   _____   ☐ Addictive USD   _____
☐ Ded. Calendar Days   _____   ☐ Deductive USD   _____
☐ No Change   _____   ☐ None USD   _____

FORM: CA-15a (Page 2 of 2)

Part III Engineer's Use

| To:
 _____
 Representative, EMPLOYER | From:
 _____
 Team Leader, Consultant | Date: |

Please find below our evaluation of the estimate submitted by the contractor for the change described above.

| Contractor's Proposed Amount
 ADD ☐ DED ☐
 USD _____ USD _____ | Negotiated/Evaluated Amount
 ADD ☐ DED ☐
 USD _____ USD _____ |

Remarks:
_____
_____
_____

| Contractor's Proposed Time Change
 ADD ☐ DED ☐
 Cal. Days _____ Cal. Days _____ | Negotiated/Evaluated Time Change
 ADD ☐ DED ☐
 Cal. Days _____ Cal. Days _____ |

Remarks:
No Additional Contract Time _____
_____
_____

Recommended Amount ADD ☐ DED ☐ USD _____ USD _____	Recommended Amount ADD ☐ DED ☐ USD _____ USD _____
Contract Amount to Date Prior to This Change	Period of Completion Prior to This Change
Contract Amount to Date After This Change	Period of Completion After This Change

Part IV EMPLOYER's Use

| To:
 _____
 Contractor
 Thru:
 _____
 Team Leader
 Consultant | From:
 _____
 Deputy Chairman
 EMPLOYER | Date: |

In Connection with Our Contract for the Construction of EWCIP CHUMATELETI-KHEVI Section, Package __, _____ Section, You Are Hereby Directed to Provide All Labor Materials Equipment, and Supervision Necessary to Complete the Changes Describe Above. This Will Be the Basis of Payment for Your Accomplishment to the Project. This Now Becomes RFP No._____

☐ Approved   ☐ Cancel Changed   ☐ Renegotiate

FORM: CA-16

PAYMENT CERTIFICATE

Payment Certificate No.   :
Covering Period   :
Date Prepared   :
Contract Title   :
Contractor   :

Contract Amount   :
Approved PFP's   :
Revised Contract Amount   :
Balance Amount to Date   :

Estimated Value of Works

Value of Work Accomplished to Date		Previous Value of Work Billed		Value of Work Accomplished This Period	
Local Currency (USD)	Foreign Currency (USD)	Local Currency (USD)	Foreign Currency (USD)	Local Currency (USD)	Foreign Currency (USD)

                                                 Local Currency(USD)   Foreign Currency(USD)

Less: 30% Recoupment of Advance Payment   : _____ _____
      5% Retention   : _____ _____
      10% VAT   : _____ _____
Total Deductions   : _____ _____
Net Amount Due this Payment Certificate   : _____ _____

                                               Signature   Date

Certificate Prepared By   : _____ : _____
                                               Team Leader

Reviewed By   : _____ : _____
                                               Representative, EMPLOYER

Recommended for Payment   : _____ : _____
                                               Deputy Director, EMPLOYER

Approved for Payment   : _____ : _____
                                               Deputy Chairman, EMPLOYER

FORM: CA-17

BILLING REQUEST FORM

Contractor : _____ Date Filed: _____
Billing Period : _____
Type of Billing(for Mobilization, Advance Payment, Progress Billing, etc.)

For Progress Billing

_____

Local Currency(USD)

Original Contract Amount : _____ _____
Value of Work Accomplished
to Date Amount of Last Billing : _____ _____
Percent(%) Accomplishment
of this Date : _____ _____
Prepared and Submitted by : _____
                                            Team Leader, Engineer

Reviewed and Endorsed by : _____
                                            Deputy Director, EMPLOYER

FORM: CA-18

SUMMARY OF WORK ACCOMPLISHED

Summary of Work Accomplished
Statement of Work Accomplished and Percentage for Time Elapsed as of
_____

	Local Currency (USD)	Foreign Currency (USD)

1. Original Contract Price
2. Revised Contract Price Due to Request for Proposal(RFP's)
3. Original Contract Time
4. Date of Contract Effectivity
5. Date of Original Contract Expiration
6. Revised Contract Time Due to Time Extension, RFP's(Variation Orders)
 	
  a. Time Extension   : _____
  b. RFP's   : _____
  c. Total Time Extension   : _____
 	
7. No. of Days Suspended due to(reasons such as: unfavorable weather condition, tendency problems, etc.)
8. Revised Contract Time
9. Revised expiry Date due to Time Extension, RFP's(Variation Order) and Work Suspension
10. Contract Time Used to Date		

Project Status	**Percentage(%)**	
1. Percentage of Contract Time Used to Date 2. Percentage of Previous Work Accomplishment 3. Percentage of Target Work Accomplished to Date 4. Percentage of Actual Work Accomplished to Date 5. Slippage		

Submitted by: Certified Correct:

_____   _____
      Team Leader   Representative
        Engineer   EMPLOYER

D.
Quality Control and Geotechnical Investigation Forms

FORM: QC-1

CERTIFICATE OF QUALITY OF TEST

Date: _____

Contract Package No. _____
Contractor: _____

    We hereby certify that we have conducted an inspection, verification and testing of materials delivered on site on the above-mentioned project.
    We further certify that we have found that the quality of materials in the following items of work corresponding to the quantity accomplished conforms with the Standard Specifications and Special Provisions of the Contract, viz:

Item No.	Description	Quantity	Accomplished
_____	_____	_____	_____
_____	_____	_____	_____
_____	_____	_____	_____
_____	_____	_____	_____
_____	_____	_____	_____
_____	_____	_____	_____
_____	_____	_____	_____

_____   _____
   Quality Assurance Engineer   Quality Assurance Engineer
          Contractor   Consultant

_____
        Team Leader

FORM: QC-2

SOUNDNESS TEST OF AGGREGATES, BY USE OF SODIUM SULFATE AASHTO, T104

Consultant : _____
Spec's Item No. : _____
Contractor : _____
Contract No. : _____
Supplied By : _____
Date Tested : _____

Sieve Size	Grading of Original Sample (%)	Weight of Test Fraction Before Test (gm)	Weight of Test Fraction After Test (gm)	Weight loss (gm)	% Loss	Weighted Percentage Loss	Maximum requirement

_____   _____
Laboratory Technician Consultant   Quality Assurance Engineer Contractor

_____   _____
Quality Assurance Engineer Consultant   Team Leader

FORM: QC-3 (Page 1 of 2)

CALIFORNIA BEARING RATIO
AASHTO T-193

Consultant		Contractor		
Station Number:	TP Number:	Sample No.:		Depth(m):
Soil Class:	Description/Item of Work			

| Description | Unit | No. of Blows | | |
		65	30	10
Weight of Wet Sample+Mold	g			
Weight of Mold	g			
Weight of Wet Sample	g			
Weight of Mold	g			
Density	g/cc			
Weight of Wet Sample+Tin	g			
Weight of Dry Sample+Tin	g			
Weight of Tin	g			
Weight of Moisture Content	g			
Moisture Content	%			
Density	g/cc			

CBR Value	Swell Test			
	Height of Sample(cm)			
	Initial Reading			
	Final Reading			
	Swell(Decimal)			
	Percent Swell(%)			
	Total Hours Soaked			
	CBR Value			
	MDD	(G/CC)		
	CBR Value@(%)		95% MDD	100% MDD
	Resistance@2.54mm			
	Resistance@5.08mm			

Date Sampled   : _____
Sampled at   : _____
Sampled by   : _____

Certified by:

_____   _____
Laboratory Technician Consultant   Quality Assurance Engineer Contractor

_____   _____
Quality Assurance Engineer Consultant   Team Leader

FORM: QC-3 (Page 2 of 2)

Station Number:		TP Number:			Sample No.:			Depth(m):		
Soil Class:		Description/Item of Work								
Number of Blows		65			30			10		
Penetration		Dial Reading	Load Kg	Stress Mpa	Dial Reading	Load Kg	Stress Mpa	Dial Reading	Load Kg	Stress Mpa
Inch	mm									
0.000	0									
0.025	0.64									
0.050	1.27									
0.075	1.91									
0.100	2.54									
0.125	3.18									
0.150	3.70									
0.175	4.45									
0.200	5.08									
0.225	5.72									
0.250	6.35									
0.275	6.99									
0.300	7.62									

Piston X-Sectional Are:
Proving Ring Constant:

65 Blows
Penetration Resistance@2.54mm x _____
Penetration Resistance@5.08mm x _____

30 Blows
Penetration Resistance@2.54mm x _____
Penetration Resistance@5.08mm x _____

10 Blows
Penetration Resistance@2.54mm x _____
Penetration Resistance@5.08mm x _____

Soaked CBR Values

Penetration Resistance		Corr.CBR Values
_____	=	_____
_____	=	_____
_____	=	_____
_____	=	_____
_____	=	_____
_____	=	_____

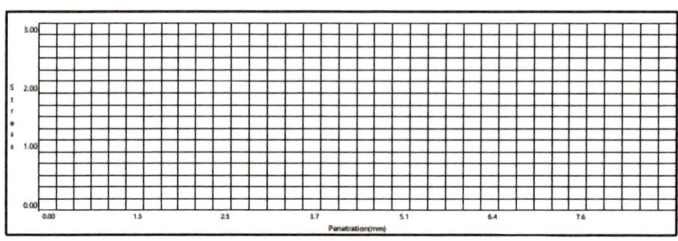

Remarks: _____

Certified by:

_____   _____
Laboratory Technician Consultant   Quality Assurance Engineer Contractor

_____   _____
Quality Assurance Engineer Consultant   Team Leader

FORM: QC-4

COMPACTION TEST FOR MDD, AASHTO T-99 OR AASHTO T-180

Consultant   : _____
Contractor   : _____
Sampled at   : _____
Sampled by   : _____
Soil Description   : _____   Contract No.: _____
Test Performed by   : _____   Sample No.: _____
Standard Test   : _____
Method Used   : _____   Date Tested: _____
Item of Work   : _____

Water Content Determination

Trail No.	1	2	3	4	5
Moisture Can No.					
Wt. of Can+Wet Soil(gms)					
Wt. of Can+Dry Soil(gms)					
Wt. of Water (gms)					
Wt. of Can (gms)					
Wt. of Dry Soil (gms)					
Water Content(%)					
Average Water Content(%)					

Density Determination

Volume of Mold(gms)					
Wt. of Moist Soil+Mold(gms)					
Wt. of Mold(gms)					
Wt. of Moist Soil in Mold(gms)					
Wet Density(gms)					
Dry Density(gms)					

Optimum Moisture Content _____ %
Maximum Dry Density _____ g/cc

Dry Density

Water Content, W(%)

_____   _____
Quality Assurance Engineer Contractor   Laboratory Technician Consultant

_____   _____
Quality Assurance Engineer Consultant   Team Leader

FORM: QC-5

FIELD DENSITY TEST, AASHTO T191
(Sand Cone Method)

Consultant   : _____   Date   : _____
Contractor   : _____   Contract No. : _____
Location   : _____   Item No.   : _____

Test Pit Number				
Station/Reference				
Sand Cone Apparatus Number				
Depth Taken				
Wt. of App. Filled W/ Sand, gms				
Wt. of App. & Remaining Sand, gms				
Wt. of Sand in Hold, Plate & Cone, gms				
Wt. of Sand in Hold, Plate & Cone				
Bulk Density of Sand, g/cc				
Volume of Test Hold, cc				
Wt. of Test Hold, gms				
Wt. of Soil from Hold, gms				
Wet Density, Kg/m^3				
Wt. of Wet Soil + Can, gms				
Wt. of Dry Soil + Can, gms				
Wt. of Water in Soil, gms				
Wt. of Dry Soil, gms				
Moisture Content, %				
Dry Density, Kg/m^3				
Lab. Max Dry Density, Kg/m^3				
Lab. Optimum Moisture, %				
Compaction				

_____   _____
Quality Assurance Engineer   Laboratory Technician
        Contractor   Consultant

_____   _____
Quality Assurance Engineer   Team Leader
        Consultant

FORM: QC-6

SPECIFIC GRAVITY AND ABSORPTION FOR COARSE/FINE AGGREGATES, AASHTO T-85/T-84

Consultant   : _____   Laboratory No. : _____
Description   : _____   Date Sample   : _____
Propose Use   : _____   Date Tested   : _____
Sampled By   : _____
Computed By   : _____

I. Coarse Aggregates
   Original Source   : _____
   Maximum Size of Aggregate   : _____

		Trial 1	Trial 2	Trial 3
1	Weight of Oven Dry Sample in Air, grams = W1			
2	Weight of Saturated Dry Sample in Air, grams = W2			
3	Weight of Saturated Sample in Water, grams = W3			
4	Absorption = $\frac{(W2 - W1)}{W1} \times 100$			
5	Bulk Specific Gravity (Oven Dry) = $\frac{W1}{W2 - W3}$			
6	Bulk Specific Gravity (SSD) = $\frac{W2}{W2 - W3}$			
7	Apparent Specific Gravity (SSD) = $\frac{W1}{W1 - W3}$			

II. Fine Aggregates
    Original Source   : _____
    Maximum Size of Aggregate   : _____

		Trial 1	Trial 2	Trial 3
1	Weight of Oven Dry Sample in Air, grams = W1			
2	Weight of Saturated Dry Sample in Air, grams = W2			
3	Weight of Saturated Sample in Water, grams = W3			
4	Absorption = $\frac{(W2 - W1)}{W1} \times 100$			
5	Bulk Specific Gravity (Oven Dry) = $\frac{W1}{W2 - W3}$			
6	Bulk Specific Gravity (SSD) = $\frac{W2}{W2 - W3}$			
7	Apparent Specific Gravity (SSD) = $\frac{W1}{W1 - W3}$			

Remarks:
_____

Submitted by: Checked by: Approved by:

_____   _____   _____
Quality Assurance Engineer Quality Assurance Engineer   Team Leader
      Contractor   Consultant

FORM: QC-7

WORKSHEET FOR LIQUID & PLASTIC TEST, AASHTO T-89/T-90

Consultant   : _____   Contractor: _____
Package No.   : _____
                      (Segment No.)   (Road Section)   (Province)
Kind of Materials : _____
Item No.   : _____
Sampled at   : _____
Original Source   : _____
                      (Name of Designation)   (Office)   (Date)
Computed by   : _____
                      (Name of Designation)   (Office)   (Date)
Lab. No. : _____

Determination	Liquid Limit						Plastic Limit	
Determination No.	1	2	3	4	5	6	1	2
Container Number								
Container and Wet Soil, g								
Container and Dry Soil, g								
Moisture Loss, g								
Container, g								
Dry Soil, g								
Moisture Content, %								
Number Blows								

Flow Curve

(x-axis: 5 10 15 20 25 30 35 40 45 50 — Moisture Contents(%))

Sieve Sizes   %

No.4   _____
No.10 _____
No.40 _____
No.100 _____

Liquid Limit _____
Plastic Limit _____
Plasticity Index _____
Group Classification _____
Group Index

Remarks
Submitted by: Checked by: Approved by:

_____   _____   _____
Quality Assurance Engineer   Quality Assurance Engineer   Team Leader
       Contractor   Consultant

FORM: QC-8

UNIT WEIGHT/MASS DETERMINATION COARSE/FINE AGGREGATES, AASHTO NO.T-19

Consultant   : _____   Laboratory No. : _____
Description   : _____   Date Sample   : _____
Propose Use   : _____   Date Tested   : _____
Sampled By   : _____
Computed By   : _____
Original Source   : _____

	Description	Loose		Rodded	
1	Container + sample, kg				
2	Container, kg				
3	Sample, kg (1) - (2)				
4	Vol. of Container, cu.m.				
5	Unit Weight/mass, kg/cu.m. (3)/(4)				
6	Average				

Remarks: _____
_____

Prepared by: Checked by:

_____   _____
Quality Assurance Engineer Contractor   Quality Control Engineer

Witnessed by: Attested by:

_____   _____
Quality Assurance Engineer Consultant   Team Leader

FORM: QC-9

WORKSHEET FOR SIEVE ANALYSIS, AASHTO T-27

Consultant: _____ Contractor: _____
Project: _____
        (Segment No.)   (Road Section)   (Province)
Kind of Material: _____ Spec's Item No.: _____
Item No.: _____
Sampled at: _____
Original Source: _____
Sampled by: _____
        (Name and Designation) (Office)   (Date)
Tested by: _____
        (Name and Designation) (Office)   (Date)
Computed by: _____
        (Name and Designation) (Office)   (Date)

Laboratory No. : _____ Wt. of Sample in Grams: _____
Dry Unit Wt. Ref. Loose: _____ Original: _____
Rodded : _____ Oven Dry: _____
Moisture Content % : _____ Wish Oven Dry: _____
Fineness Modules : _____

Sieve Size	Weight Retained	% Retained	Commulative			Specs Requirement	Remarks
			Wt. Passing	% Passing	% Retained		
3							
2 1/2							
2							
1 1/2							
1							
4-Mar							
2-Jan							
8-Mar							
No.4							
No.8							
No.10							
No.16							
No.30							
No.40							
No.50							
No.100							
No.200							
Pan							
Wash(-)200							
Total							

Remarks/Recommendation
  Submitted by:
  _____
    Quality Assurance Engineer Contractor   Quality Assurance Engineer Consultant
  Verified by:
  _____
    Team Leader

FORM: QC-10

SUB BASE CONTROL

Consultant : _____
Contractor : _____   Contract No.: _____
Location : _____   Date: _____
Station : From Km. _____   to Km. _____
Source : _____   Material: _____

Tolerance from the design	Required	Measured	Permitted Variation			
Layer Thickness						
Elevation of Surface						
Surface Irregularity						
Crossfall						
Longitudinal grade over 25m						

Test Result		Required Criteria	Samples				
			1	2	3	4	5
Gradation	% Passing 50 mm						
	25 mm						
	9.5 mm						
	0.425 mm						
	0.075 mm						
LL, %							
PI. %							
LLA, %							
MDD, T-180 g/cm³							
CBR, %							
OMC, %							
FDT, g/cm³							
Compaction: 100FDT/MDD							

The fraction passing the 0.075mm sieve shall not be greater than two thirds of the fraction passing 0.425mm sieve.

_____   _____
Quality Assurance Engineer   Laboratory Technician
Contractor   Consultant

_____   _____
Quality Assurance Engineer   Team Leader
Consultant

FORM: QC-11

BASE CONTROL

Consultant   : _____
Contractor   : _____   Contract No.: _____
Location   : _____   Date: _____
Station   : From Km. _____   to Km. _____
Source   : _____   Material: _____

Tolerance from the design	Required	Measured	Permitted Variation
Layer Thickness			
Elevation of Surface			
Surface Irregularity (by 3m straight edge)			
Crossfall			
Longitudinal grade over 25m			

Test Result		Required Criteria	Samples				
			1	2	3	4	5
Gradation	% Passing 37.50 mm						
	25.00 mm						
	19.00 mm						
	12.45 mm						
	4.750 mm						
	0.425 mm						
	0.075 mm						
LL,	%						
PI.	%						
LLA,	%						
MDD, T-180	g/cm³						
Material,<19mm CBR %							
OMC,	%						
FDT,	g/cm³						
Compaction: 100FDT/MDD							

The fraction passing the 0.075mm sieve shall not be greater than two thirds of the fraction passing 0.425mm sieve.

_____   _____
Quality Assurance Engineer   Laboratory Technician
Contractor   Consultant

_____   _____
Quality Assurance Engineer   Team Leader
Consultant

FORM: QC-12

CONCRETE POURING FORM

Station Covered : _____
Structure : _____
Total Volume : _____
Item No. : _____

Truck No.	Volume	Start Mixing	Left C.B.P.	Arrived at Site	Start of Placing	Completed Placing	Temperature		Slump, mm	Yield, m3	Remarks
							Mix	Ambient			

Prepared by: Checked by: Approved by:

Quality Assurance Engineer   Quality Assurance Engineer   Team Leader
Contractor   Consultant

FORM: QC-13

CONCRETE POURING REPORT

Structure/Segment : _____
Station : _____

Time of Pouring : _____   Weather : _____
Class of Concrete : _____   Part of Structure Poured : _____
Method of Mixing : _____   Mixer Capacity : _____
CEMENT   Designed Cement Factor : _____
Brand : _____   Computed Volume : _____
Condition of Storage : _____

Actual Quantities Per Batch

Cement   : ____ bags   Water : ____ lit.
Fine Aggregates   : ____ kg/m³
Coarse Aggregates   : ____ kg/m³
Cement Used (Total) : ____ kgs. Vol. of Conc. Poured ____ cu.m.

	Specification REQTS.	Actual
Cement Factor bags/cu.m.	: _____	_____
Water, lits	: _____	_____
Slump, mm.	: _____	_____

Kind of Sample   : _____   Sample Identification : _____
Quantity Represented : _____   Method of Curing   : _____
Sampled by   : _____
                        : _____
                              (Name and Designation)

Submitted by: Reviewed by: Verified by:

_____   _____   _____
Quality Assurance   Quality Assurance   Team Leader
Engineer Contractor   Engineer Consultant

FORM: QC-14

WORKSHEET FOR COMPRESSIVE STRENGTH TEST OF CONCRETE, AASHTO T-22 FOR FLEXURAL STRENGTH TEST, AASHTO T-97/T-177

Date   : _____   Station: _____
Item Number   : _____
Structure   : _____
Part of Structure   : _____ Cement Factor: _____ kg/cu.m.

Laboratory No.   : _____
Date Sampled   : _____
Date Tested   : _____
Tested by   : _____
Age in Days   : _____
Type of Test   : _____
Class of Concrete   : _____
Specifications   : _____
Brand of Cement   : _____

				Average
Sample No.(ID)				
Slump (mm)				
Wt. of Cylinder (KGS)				
Machine Reading (KGS)				
Strength (PSI)				

Remarks / Recommendations _____

Submitted by: Checked by: Approved by: Noted by:

_____   _____   _____   _____
Lab, Tech   QA, Contractor   QA, Consultant   Team Leader

Appendix

FORM: QC-15

PC CONCRETE QUALITY CONTROL

Engineer   : _____
Contractor   : _____ Contract No.: _____
Structure/Placement: _____ Station: _____
Concrete Content   : _____ Part of Structure: _____
Sample Date   : _____ Time: _____ Weather: _____
Aggregate Tested   : ☐ No ☐ Yes Test No.: _____ Source: _____
Cement Brand   : _____ Method of Mixing: _____

Constants and Formula					
			Concrete Beam, T97:		
a. Diameter	:		a. Average Width	:	
b. Length	:		b. Average Depth	:	
c. X-Sect Area	:		c. Span Length	:	
d. COMP. STR	=	$\dfrac{Breaking\ Load}{Cross-Section-Area}$	d. If break within Middle Third	:	
			Flex. Str. = PL/bd^2	:	
			e. If break outside Middle Third	:	
			Flex. Str. = $3Pa/bd^2$	:	

Quality Requirements					
Strength		Slump	Conc, Mix	Remarks:	
Compressive	Flexural				

Tests on Fresh Concrete, T119, T121 and T196:

Temp. of Air	Temp. of Conc.	Slump		Air Voids	Batch for Strength Test:
		Test Run	Ave.		Specimen
					a. Cylinder/Beam(circle one)
					b. Number of specimen Taken _____ .
					c. Identification No. of each specimen

Test on Hardened Concrete, T22 of T97:

Identification Number of Specimen	Unit Wt. (Kg.)	No. of Spec. in Batch	Date Cast	Age Days	Strength Test				Remarks
					Compressive		Flexural		
					Breaking Load(Kg.)	Strength (Kg./cm²)	Breaking Load(Kg.)	Strength (Kg./cm²)	

Comment of Type of Fracture and Any Specimen Defects, etc:
Technician   : _____ Date: _____ For: Slump/Specimen Taken/
Quality Assurance Engineer: _____ Date: _____ Compressive Break/
Resident Engineer   : _____ Date: _____ Flexural Break

_____   _____
Quality Assurance Engineer   Laboratory Technician
       Contractor   Consultant

_____   _____   _____
Quality Assurance Engineer   Team Leader   Representative
       Consultant   EMPLOYER

FORM: QC-16

BITUMINOUS PLANT MIX CONTROL

Consultant   : _____

Contractor   : _____   Contract No.   : _____

Location   : _____   Date Tested   : _____

Sieve Size		Job-mix formula	Measured Results	Range	Tolerances from job-mix
mm	In/No.				
12.7	1/2				
9.5	3/8				
4.75	No. 4				
2.36	No. 8				
1.18	No. 16				
0.600	No. 30				
0.300	No. 50				
0.150	No. 100				
0.075	No. 200				
Sand Equivalent					
LA Abrasion					
Bulk Spec Gravity					
Stripping					

Mineral Filler:
    Min. one (1) Gradation and Plasticity Test for every 75m³ or fraction thereof

0.600	No. 30				
0.300	No. 50				
0.075	No. 200				
PI					

Bituminous material: Min. one (1) penetration test for every 40ton.

Penetration					

Mix:
    Min. one (1) Grading, Extraction, Stability and Compaction Test for every 75 m³ or fraction thereof.

Bitumen Content, %					
Density, g/cm³					
Air Voids, %					
Stability, lb (corrected)					
Flow X 0.01 in.					
Voids in Mineral Aggregate, %					
Specimen Height, in					
Bulk spec. Gravity					
Temperature of Mix					

_____   _____
Quality Assurance Engineer Contractor   Laboratory Technician Consultant

_____   _____
Quality Assurance Engineer Consultant   Team Leader

FORM: QC-17

BITUMINOUS CONCRETE SURFACE COURSE CONTROL

Consultant : _____ Item No. : _____
Contractor : _____ Contract No. : _____
Station : _____ Lane : _____
Surface Width : _____
Course/Pavement Thickness: _____

Bit. Conc. Type	Laboratory	Field
Specimen No.		
Date Sampled		
Date Tested		
Average Thickness		
(A) Specimen Wt. in Air, g		
(B) Specimen Wt. in H$_2$O, g		
(C) Saturated Surface Dry Specimen weight, g		
(D) Bulk Specific Gravity $D = A / (C - B)$, g/cm^3		
(E) Maximum S. Gravity AC by wt. of agg., %		
Air Voids $n = [(E - D) / E] \times 100$, %		
Degree Compaction= (D Field / D Lab) × 100%		
Marshall Stability, lb (adjusted)		
Flow, 0.01 in.		
Surface Irregularity, mm		
Specifications Min. Degree of Compaction: _____ AC Range (% by wt.): _____ Min. Marshall Stability: _____ Surface Tolerance (Non Straight edge): _____ Flow Range: _____		

_____   _____
Quality Assurance Engineer   Laboratory Technician
Contractor   Consultant

_____   _____
Quality Assurance Engineer   Team Leader
Representative Consultant

FORM: QC-18

ORGANIC CONTENTS

Engineer		Contractor					
Test Pit Number							
Station Number							
Sample Number			1	2	1	2	3
Depth	m						
Weight of Dry Soil (4 hours)	gm						
Weight of Dry Soil After Combustion (24 hours)	gm						
Weight Loss	gm						
Organic Content	%						

Submitted by: Remarks _____
_____
_____

Certified by

_____   _____
Laboratory Technician   Quality Assurance Engineer
Consultant   Consultant

Approved by

_____   _____
Team Leader   Date

E.
Environmental Monitoring Forms

FORM: EM-1

CERTIFICATE OF LABORATORY ANALYSIS

Name of Source : _____

Laboratory Receipt Number :_____   Date Sampled : _____

Certificate Number   : _____   Date Received : _____

Date Reported   : _____   Date Analyzed : _____

Source of Sample   : _____

Station Number	TSP μg/ncm	SO2 μg/ncm	NO2 μg/ncm
Detection Limit			
Method			

Analyzed By: Certified By:

_____   _____
         Analyst   Laboratory Head

Verified By: Noted By:

_____   _____
         Engineer   Engineer

Appendix   205

FORM: EM-2

RESULTS OF PHYSICAL AND CHEMICAL ANALYSIS FOR WATER

Source : _____
Address: _____
Date Sampled : _____   Sampled By : _____
Date Received : _____   Date Completed : _____
Date Analyzed : _____   Date Reported : _____

Sample No.	Sample Identification	Station Number	Laboratory Number

Parameters	Stations (Specify Station Designation Based on EMP)				
pH					
Color					
Dissolved Oxygen					
BOD (5 Day, 20°C), mg/L					
Total Suspended Solids, mg/L					
Oil and Grease, mg/L					

Remarks: _____
_____

Standard methods for the examination of water is referred to 18th Edition of 1995, APHA, AWWA, WEF

Analyzed By: Certified By:

_____   _____
        Contractor   Consultant

                                     Noted By:

                                     _____
                                              EMPLOYER

FORM: EM-3

AMBIENT AIR QUALITY MONITORING DATA

Source: _____   Address: _____

Station	Location	Date	Time	TSP	SO^2	NO^2	Remarks

Note: TSP, SO^2 and NO^2 values are given in mg/Ncum or micrograms per cubic meter at 25 degrees and 1 atm.

Recorded by: Certified by: Noted by:

_____   _____   _____
Contractor   Consultant   EMPLOYER

FORM: EM-4

NOISE LEVEL MONITORING DATA

Source: _____ Address: _____

Station Number	Location	Date	Time	Period	Noise Level					DENR Standard			
					Class AA	Class A	Class B	Class C	Class D	Class AA	Class A	Class B	Class C	Class D

Recorded by: Checked by: Noted by:

_____ _____ _____
Contractor   Consultant   EMPLOYER

FORM: EM-5

WATER QUALITY MONITORING DATA

Source: _____ Address: _____

Station Number	Location	Date	Time	BOD mg/l	TSS mg/l	O.G. mg/l	DO mg/l	pH
DENR Water Quality Criteria (DAO 34)		Class A		5	50	1	70%**	6.5-8.5
		Class B		5	>30%*	1	70%**	6.5-8.5
		Class C		7	>30%*	2	60%**	6.5-8.5
		Class D		10	>60%*	5	40%**	6.0-9.0

Note:
* Per-cent Increase
** Per-cent of saturation value, i.e., for water at 30 degrees centigrade the DO saturation value is 7.63 mg/l

** 70% saturation value at 25 degrees Centigrade = 5.34 mg/l
** 60% saturation value at 25 degrees Centigrade = 4.58 mg/l
** 40% saturation value at 25 degrees Centigrade = 3.05 mg/l

Recorded by: _____ Checked by: _____ Noted by: _____

Contractor Consultant EMPLOYER

Appendix

닫는 글

건설 산업은

　우리나라의 건설 산업은 세계적인 설계 및 시공 기준에 따라 고속도로, 교량, 터널, 철도, 공항, 항만, 수자원, 신도시, 산업단지, 초고층 건물 등 다양한 분야에서 성과를 이뤄냈다. 발주기관은 Project Life Cycle에 따라 예산과 정책이 뒷받침되었고, 건설회사와 엔지니어들은 신기술 개발과 첨단 공법의 도입을 통해 설계 및 시공 능력을 키워 왔다.

건설사업관리는

　우리나라의 건설 사업관리는 발주기관이 기획-설계-시공-유지관리 전 과정을 주도하여 성과를 이루었다. PM 방식의 사업사례가 거의 없고 공사 관리는 시공 단계에서 관련 법령에 따라 시행되고 있다.

지금 건설사업에서의 PM은

　건설사업관리는 사업관리회사(Project Management Company)의 Project Manager(PMr)가 발주기관으로부터 책임과 권한을 위임받아 Project Life Cycle 전체를 운영하는 발주기관의 파트너로서 사업 목표 달성을 위하여 활동하는 것이다.

시공관리는 사업의 규모와 공사 종류에 맞추어 투입되는 부분 별 엔지니어의 조직을 말한다. 이는 공사 관리회사(Construction Management Company)의 Construction Manager(CMr)가 발주자를 대신하여 공정, 품질, 안전 등을 담당하는 것이다.

그러나 현실에서는 PM이 실현되기 어렵다. 당초에 PM 사업을 평가할 만한 사례가 없다. 또한 PM 사업도 발주기관이 직접 사업관리를 하거나 PM은 단지 "공사관리를 확대"한 개념으로 발주기관이 대부분의 권한을 쥐고 있다. 이는 개발시대에서는 효과적일 수 있다.

이제 건설 사업은 대형화, 전문화되어 복잡한 사업이다. 관련기관과의 협의, 신설 또는 기존 시설의 증설, 민원의 조치 등 많이 변화되고 있다. 사업주체자와 발주기관의 업무가 한 단계 높아져야 한다.

발주기관은

발주기관은 본연의 정책기획 및 사업관리 감독 업무에 집중해야 한다. 발주자는 공통된 목적이나 목표를 달성하기 위해 상호 연관된 단수 또는 복수의 프로젝트들을 조정하고 관리한다. 수개의 프로젝트를 통합적으로 기획, 관리, 조정, 승인, 감독하는 사업의 연속발주자[1](Serial Builder)이다. 즉 Program Management체계이며 프로젝트들의 관리자(Project of Projects)이다.

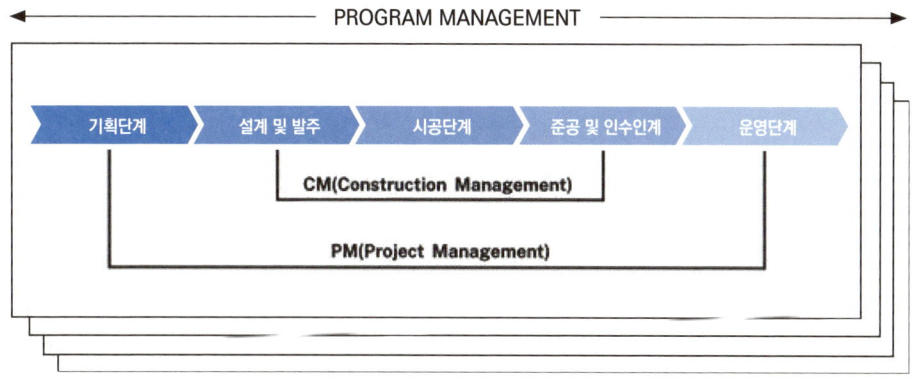

---
1) Program Management, Chuck Thomsen

발주기관은 대규모 인프라 개발, 도시 재개발, 교통망 구축, 복합 개발사업 등에서 PgM 체계의 업무가 된다. PM은 단일 프로젝트의 대상으로 사업분석, 예산, 계약, 리스크를 관리하고 PgM은 복수의 연계된 프로젝트를 대상으로 자원을 최적화하여 각 프로젝트 간의 우선순위를 조정하고 통합된 목적을 달성하는 데 있다.

예를 들면 PM은 하나의 고속도로 건설을 관리하고 PgM 체계에서 전국 고속도로망 구축 프로그램 관리하는데 프로젝트 간 충돌을 예방하고 성과를 모니터링하여 보고 체계를 완성하는 것이다. PM은 신도시 개발 택지조성, 도로, 상하수도, 학교, 병원 등의 단일 프로젝트를 관리한다. PgM은 수십 개 프로젝트를 통합관리하는 신도시, 국가 기간교통망 사업, 고속철도, 고속도로, 환승센터 등 각기 다른 프로젝트를 연계하여 관리한다. 올림픽·엑스포를 개최하면 경기장, 도로, 숙박시설, 공항 확장 등 다수 프로젝트를 조정하는 것이다.

발주기관은 PgM 체계로 통합형 건설관리 방식으로 정착되어야 한다. 건설 사업관리(PM), 공사관리(CM)를 넘는 총괄 지휘 체계로써 매우 중요한 역할이다.

계약 방식은 본질이 아니다

계약 방식은 건설사업관리가 아니다. 계약 방식은 시대나 사업 특성에 따라 달라질 수 있다. 턴키(Turn-key), EPC(Engineering Procurement Construction), PPP(Public Private Partnership), PF(Project Financing) 등 다양한 형태의 계약이 있다. Turn-key는 설계·시공 일괄 발주하고 EPC는 엔지니어링·구매·시공 일괄 수행하며 PPP/PF는 민간 자본을 활용한 방식이다. 이것이 사업의 성공을 보장하지는 않는다. 사업의 성공을 좌우하는 것은 계약의 방식이 아니라 '누가 어떻게 관리하는가?'이다. PM은 또 하나의 계약 방식이 아니다.

건설 사업은 PM으로

PM은 사업의 시간과 비용을 명확히 정의하는 데서 출발한다. PM 방식은 단순한 공사 관리를 넘어 사업의 경제성과 타당성을 객관적으로 평가하여 실행 여부를 결정하는 기준이 된다. 또한 사업이 설계도와 기술시방서에 따라 정확히 수행될 수 있도록 공사의 시간과 비용을 실질적으로 검증해 주는 역할을 한다.

PM은 신속하고 합리적인 의사결정을 할 수 있도록 정보를 제공한다. 연속성과 일관성을 갖고 수행할 수 있다. 발주기관의 직간접적인 영향을 최소화하여 균형을 맞추어 자율적이고 책임 있는 문화를 정착시킬 수 있다. 사업의 투명성을 높이고 불필요한 갈등을 줄이며 협력 중심의 사업 수행을 가능케 한다.

PM 방식으로 공사 기간을 지키고 안전사고 없이 최소의 비용으로 가능하게 만들어야 한다. 사업 주체자와 발주기관 그리고 우리 모두의 목표가 이루어질 수 있다.

젊은 엔지니어가 오고 있다.

그들은 Smart IT 기술을 갖추었고 영어로 말하고 쓰는 능력이 높다. 현지 문화에 능숙하고 디지털 건설 산업 현장의 주역이다. 그들은 Robot, Autonomous Vehicle, Artificial Intelligence, Big Data, Internet of Things, Virtual Reality, Augmented Reality, 3D Printers, Drones, Cloud, Global Positioning System, Building Information Modeling, CCTV, Web Camera, Monitor 등을 활용하여 PM과 IT 기술을 융합시켜 Smart PM으로 운용할 수 있다.

AI는 수요 예측과 입지 분석을 정밀하게 수행하고, Drone 촬영으로 측량하고 지형도를 만들어 노선을 비교하고, 용지 도면을 만들어 지장 물건을 확인하고, 해당 토지 정보를 분석하고, 구조물 위치를 최적화하고, 노선별 기본계획을 만들어 준다. Drone과 Robot은 고위험 작업을 대신한다. BIM과 Cloud 플랫폼은 설계자, 시공자, 발주자가 실시간으로 협업할 수 있게 하며, IoT 센서는 구조물 상태를 실시간 안전과 품질을 모니터링 하며 공사 물량과 공정을 비교하며 시공할 수 있다. VR, AR, AV, 3D 기술은 현실과 가상을 연결하여 설계, 시공, 유지관리의 업무를 입체적으로 판단할 수 있게 해준다. 이제는 정보와 데이터, 예측과 통합의 기술 산업으로서의 건설을 이야기해야 한다.

PM 방식은 이러한 조직혁신과 기술 융합을 가능하게 한다. PM은 단순한 관리 도구가 아니라 스마트 건설의 실행 시스템이다. PM 방식을 통해 젊은 엔지니어들이 더 많이 프로젝트에 참여할 수 있도록 해야 한다.

PgM으로 'One Team Korea'로 세계 무대로 나서야

그동안 발주기관은 국토종합개발계획으로 SOC사업을 기획하고 건설하여 국가 발전에 견인 역할을 하였다. 이것이 우리나라 건설 산업의 소중한 자산이다. 건설사업관리를 시스템으로 만들 수 있다.

국책사업이나 민간투자개발사업에서 축적된 건설 사업관리 경험한 엔지니어들은 해외 개발 MDB 사업, 재정지원 사업에서도 경쟁력을 발휘할 수 있다. 이들은 국내 건설공사에서 국제적인 설계기준에 따라 시공한 경험은 해외공사의 경력으로 대체할 수 있다. 오히려 건설공사의 규모나 공법, 기술은 국내 공사가 앞서있다. 해외공사가 아니라는 이유로 해외 건설 사업에 참여가 제한되는 것은 불공평하다. 상대국의 정부와 협약 등으로 해결되어야 한다.

발주기관은 엔지니어(PMr, CMr)들이 민간 조직의 위상을 제도화할 수 있도록 국제 기준에 따라 업무 절차와 책임을 관련 법으로 제정되어야 한다. 발주기관은 PM, CM의 시스템이 정상 활동이 되도록 지휘하고 관리해야 한다.

현재 해외공사는 PM 방식 또는 설계·시공 통합형 발주 체계로 형성되고 있다. 국내에서 경험한 엔지니어들과 One Team Korea로 공공기관, 건설 사업 관리회사, 설계회사, 건설회사, 전문 건설회사, 자재 공급회사, 장비 운영회사, 금융기관이 통합된 시스템으로 참여할 수 있다.

세계는 아직도 많은 개발도상국이 사회 간접 자본(Social Overhead Capital) 시설이 필요하다. K-Engineer가 One Team Korea가 되어 세계 건설시장으로 나가야 한다.